AS Science for Public Understanding

Editors: Andrew Hunt and Robin Millar

THE UNIVERSITY *of York*

UYSEG

SCIENCE EDUCATION GROUP

AQA

The Nuffield Foundation

Heinemann

Editors: Andrew Hunt and Robin Millar.

Authors: David Applin, Paul Bowers Isaacson, Ann Fullick, Andrew Hunt, Angela Melamed, Robin Millar, Bryan Milner, Michael Reiss.

Designed and typeset by Cambridge Publishing Management
Illustrated by Geoff Ball
Printed and bound in Spain by Edelvives

Heinemann Educational Publishers
Halley Court, Jordan Hill, Oxford, OX2 8EJ
a division of Reed Educational & Professional Publishing Ltd
Heinemann is a registered trademark of Reed Educational & Professional Publishing Ltd

OXFORD MELBOURNE AUCKLAND
JOHANNESBURG BLANTYRE GABORONE
IBADAN PORTSMOUTH NH (USA) CHICAGO

© Nuffield Foundation, 2000

First published 2000

ISBN 0 435 65466 7

04 03 02 01
10 9 8 7 6 5 4 3 2

Acknowledgements
The authors and publishers would like to thank the following for permission to use photographs: Science
Photo Library: Figs. 1.1, 1.2, 1.4, 1.5, 1.10, 2.1, 2.4, 2.6, 2.7, 2.8, 2.10, 2.13, 3.1, 3.3, 3.4, 3.6, 3.7, 3.11, 3.12,
3.13, 3.17, 3.19, 4.3, 4.4, 4.5, 4.6, 5.1, 5.5, 5.6, 5.7, 5.8, 5.9, 6.1, 6.8, 6.9, 7.4, 7.5, 7.8, 7.9, 7.10, 7.12, 8.1,
8.2, 8.5, 8.8, 8.9, 8.13, 8.14, 8.16, 10.4, 10.12, 10.14, 11.1, 12.12, 12.13, 13.1, 13.2, 14.7, 14.12, 15.1, 15.2,
15.7, 16.5, 16.6, 16.8, 16.9a, 16.9b, 16.11, 16.15; Mary Evans Picture Library: Figs. 1.3, 1.9, 3.8b, 4.9, 8.7,
8.11; Institut Pasteur: Figs. 1.7, 1.8; Popperfoto: Figs. 2.11, 3.8a, 11.14; Hulton Getty: Figs. 2.12, 9.1a;
Oxford Scientific Films: Figs. 3.2, 3.5; John Olive, Dept. of Chemistry, University of York: 3.10; Corbis: Figs.
3.15, 7.13, 9.1b, 9.1c, 10.17, 10.18a, 10.18b, 11.5, 11.9; Stone: Figs. 4.1, 10.1, 14.1; Rex Interstock: Fig. 4.10;
Holt Studios: Fig. 7.7; Bill Day: Fig. 8.12; The Advertising Archives: Fig. 10.7; PA News: Fig. 10.10;
Environmental Images: Figs. 11.7, 12.1; Catalyst: The Museum of the Chemical Industry: Fig. 11.13;
Greenpeace: Fig. 12.14; NASA Figs 15.19, 16.1.

Cover photos: gas giant: Photodisc; biosphere: Corbis; acupuncture: Science Photo Library/Tim Malyon and
Paul Biddle.
Picture research by Caroline Thomas

Thanks are due to the following for permission to reproduce extracts and illustrations. HMSO: Figs. 5.2, 5.4;
Oxford University Press: Figs. 9.4, 9.20, 9.21, 10.15, 16.14; International Energy Agency: Figs. 9.9, 9.10,
9.11, 9.22; UK Government Panel on Sustainable Development: Fig. 9.14; European Commission: Fig. 9.17;
HarperCollins: Figs. 10.5, 10.6; *New Scientist*: Figs. 10.9, p. 136 q.10, 10.16, 10.19, 14.8, 14.11, 14.14;
OECD: Fig 10.11; AEA Technology website/Department of the Environment, Transport and the Regions: Figs.
11.6, 11.8; Penguin: pp. 181, 191, 206; Wolters Kluwer: Figs. 11.10, 11.11; Heinemann: Figs. 12.3, 12.4;
IPCC: Fig. 12.8; Climate Impacts Programme: Figs. 12.9, 12.10, 12.11; *The Times, The Sun, Today*: Fig.
13.11; *The Telegraph*: Fig. 13.12; NRPB: Fig. 13.14, 13.16; *The Scotsman, The Independent*: Fig. 14.2;
The Daily Express: Fig. 14.3; Longman: Fig. 15.9; Science Museum, London: 15.13; Association for Science
Education: 15.15.

The publishers have made every effort to trace the copyright holders, but if they have inadvertently overlooked
any, they will be pleased to make the necessary arrangements at the first opportunity.

Contents

Acknowledgements		*ii*
Contents		*iii*
Foreword		*iv*
Introduction		*v*

ISSUES IN THE LIFE SCIENCES — 1

Chapter 1	The germ theory of disease	1
Chapter 2	Preventing diseases	13
Chapter 3	Medicines to treat disease	26
Chapter 4	Health risks	43
Chapter 5	Alternatives in medicine	62
Chapter 6	Genetic diseases	75
Chapter 7	Genetic engineering	87
Chapter 8	Evolution: understanding who we are	101

ISSUES IN THE PHYSICAL SCIENCES — 115

Chapter 9	Using fuels	115
Chapter 10	Electricity supplies	129
Chapter 11	Air quality	143
Chapter 12	Fuels and the global environment	155
Chapter 13	Radioactivity	167
Chapter 14	Radiation risks?	179
Chapter 15	Understanding the solar system	192
Chapter 16	Understanding the universe	208

STUDY GUIDE — 221

| **Chapter 17** | Coursework guidance | 221 |
| **Chapter 18** | Revision and exam preparation | 236 |

| *Index* | | *244* |

Foreword

This AS course in *Science for Public Understanding* is a means of broadening the curriculum for those whose interests lie mainly in the arts and humanities or for giving those who study science an opportunity to reflect on their specialist interests in a broader context.

While developing the course and working on this book, we have drawn on the work and thinking of earlier projects which have explored the interplay between science, technology and society. The most influential ones in the UK were all published by the Association for Science Education: *Science in Society, Science in a Social Context* (SISCON) and *Science and Technology in Society* (SATIS).

Like the earlier projects, this course is built around the study of topical issues and key episodes in the history of science. It differs in emphasis from its precursors, however, in two important respects. First, we are much more specific about the expected learning outcomes related to the nature of science. Secondly, the understanding of science content expected is set out more explicitly, and aims not so much to extend students' understanding as to help them develop a more rounded and mature view of the major science explanations they have met during their National Curriculum studies up to the age of 16.

We have developed this AS course in *Science for Public Understanding* in close collaboration with Philip Pryor of the Assessment and Qualifications Aliance (AQA) and with two teachers, Angela Melamed and Paul Bowers Isaacson. We have also learnt much from the pilot course organised by AQA which ran in about 25 colleges and schools over the 2 years leading up to the launch of the course in September 2000. We are very grateful to the teachers who volunteered to take part and to their students who sat the pilot examinations.

We would like to thank our co-authors for their contributions to this book and for working so hard to create a new style of science text in a very short time. We are also indebted to Kay Symons and Lindsey Charles of Heinemann Education for supporting the publication of the book with great enthusiasm.

This book and the project web site have been developed through a collaboration between the University of York Science Education Group and the Nuffield Curriculum Projects Centre funded by a grant from the Nuffield Foundation.

Andrew Hunt
Robin Millar

Introduction

The three strands of the course

As you work through this course in *Science for Public Understanding* you study a series of *topical issues* and episodes from the history of science. In each topic you learn to apply your understanding of *scientific explanations*. You also reflect on the way that science itself works, and how it impacts on society, by considering a series of *ideas about science*.

The opening page for each of the first 16 chapters in this book shows you how the three strands of the course are interwoven: the issues, the science behind the issues, and what a study of the issues tells you about science and society.

Studying the course

During this course you will take part in discussions, debate issues and form your own opinions. You are not expected to recall all the information in this book. The details of the issues and case-studies are not as important as the scientific explanations and the ideas about science that lie behind them. You should focus on the questions in each chapter, which indicate the kinds of things you are expected to be able to do with the information provided, and the general ideas you are expected to be able to bring into your discussion.

The coursework allows you to keep up to date and explore your own interests. You choose which topical scientific issue you will study and you select a piece of popular science writing to read, enjoy and analyse.

How this book is organised

This book matches the structure of the specification for the AS course which consists of three modules.

Module 1: Issues in the life sciences   chapters 1–8

Module 2: Issues in the physical sciences chapters 9–16

Module 3: Coursework guidance   chapter 17

Finally chapter 18 provides advice about revision and preparation for the two examinations which test modules 1 and 2.

The specification, together with a handbook for teachers and specimen examination papers, are all available from the Awarding Body, AQA. The latest information is available on the AQA web site: (http://www.aqa.org.uk). You will find a commentary on the questions in this book on the project web site (http://www.nuffieldfoundation.org/spu) where you will also find links to other web sites with up to date and authoritative information about the issues which feature in the course.

Chapter 1

The germ theory of disease

The issues

In many parts of the world millions die each year from infectious diseases such as cholera, TB and malaria. Even in richer countries food poisoning makes many people very ill and can kill.

In some parts of the world people are too poor to ensure that their water is safe to drink or to pay for the treatments which can prevent or cure diseases. Everywhere people, even when they have been taught the germ theory of disease, ignore basic rules of hygiene and lay themselves open to the risk of infection.

Preventing and curing disease costs money, and across the world there are real issues about how money for health care is spent. In order to make the right decisions, and to fight infectious diseases whenever and wherever they occur, it is necessary to understand the causes of such diseases and how they are passed from one individual to another.

The science behind the issues

We now know that the 'germs' which cause infectious diseases are small organisms including bacteria, viruses and fungi (Figure 1.1). These micro-organisms are present in the environment and can be passed from one infected individual to another. Under ideal conditions the body defends itself against these invading germs using the immune system, but often not before unpleasant symptoms are experienced. In some cases the patient is dead before the immune system has done its work. By developing an understanding of how diseases are caused and spread we can move towards preventing or curing them – though this is not always as easy as it sounds.

What this tells us about science and society

People have known about infectious diseases for centuries. But it was only in the 19th century that they began to understand what causes them. So having information (data) is not enough to guarantee that we will find a theory which can explain things. Coming up with a new theory needs creative imagination – to see patterns in the data and think up possible explanations. More than one explanation can account for the same data – so there is always room for disagreement and people's past experiences and wider commitments can influence their judgement. A scientist has to persuade the whole scientific community of his or her idea before it becomes accepted as 'reliable knowledge'.

Figure 1.1

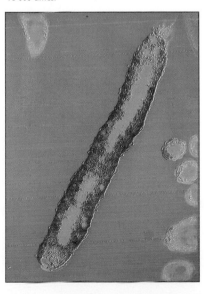

Vibrio cholerae is the bacterium which causes cholera in humans. This false-colour image from an electron microscope is magnified about 18 000 times.

Figure 1.2

A women's ward of the Bridewell hospital in London in 1808. The beds are piles of straw. In the worst hospitals at this time the death rate from puerperal fever was as high as one new mother in every three.

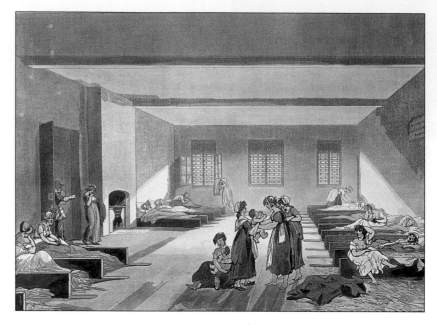

The story of Ignaz Semmelweis

After giving birth, women are vulnerable to infections, but in the days when most women had their children at home attended by their female relatives, serious infections after delivery were relatively rare. But as society moved into the late 18th and early 19th centuries medical intervention began to cost women their lives.

Doctors became increasingly involved in the delivery of babies at home, and hospitals were set up with maternity wards for the 'safe' delivery of infants (Figure 1.2). These hospitals were used mainly by the poorer women who could not afford medical care in their own homes. The women began to die a horrible death. Within days or even hours of giving birth mothers developed a range of symptoms including pain and tenderness of the abdomen, together with a rapid pulse and high fever. Severe pain, inflammation of the womb, vomiting, convulsions and death followed within five days. This dreadful illness, known as puerperal fever or childbed fever, regularly claimed the lives of about one woman in every five who gave birth to a baby. Even when women delivered at home, the fever became more common, but it depended on who delivered the baby. Some doctors lost most of their maternity patients, while others never had a case of childbed fever.

There were fierce debates about the causes of this mystery disease. In Britain, Charles White and, in America, Oliver Wendell Holmes tried hard to persuade people that it was doctors and nurses who were spreading the disease from patient to patient. However, the person we now remember for gathering the scientific evidence to show how the disease was spread was Ignaz Philipp Semmelweis, an Hungarian physician.

Ignaz Semmelweis was born in Buda, Hungary, on 1 July 1818. Soon after he qualified as a doctor he became an assistant at the maternity clinic of the Vienna General Hospital. The hospital had two delivery rooms, one staffed by female midwives and the other by

Questions

1 Infectious diseases have affected people throughout human history. Why do you think that it took so long for people to begin to investigate and discover the causes?

2 The numbers of women dying of childbed fever increased as doctors and hospitals became more involved in the delivery of babies. Using your 21st century knowledge about disease, why do you think this was the case?

3 How is your behaviour affected by your belief in the germ theory of disease? How would your behaviour be different if you did not know about bacteria and viruses?

medical students. Over 12% of the women delivered by the young doctors died of puerperal fever. This was over three times the percentage of women dying from the other delivery room. Semmelweis realised that the medical students were often dissecting a dead body as part of their training and then moving straight on to delivering a baby without washing their hands first. He wondered if they were carrying the cause of disease on their hands from the corpses to their patients.

Then a colleague of his, Jacob Kolletschka, cut himself while carrying out an autopsy and subsequently died from symptoms identical to those of puerperal fever. For Semmelweis this confirmed his idea that puerperal fever was caused by an infectious agent. He immediately insisted that his medical students wash their hands in chlorinated lime before they entered the maternity ward, and eventually he insisted that they should wash between each patient. Within 6 months the mortality rate of his patients had dropped to a quarter of the original figure, and after 2 years the death rate was down to only 1.3% of the women who delivered their babies in his wards.

Semmelweis presented his findings to other doctors. He was sure that they would recognise from his evidence that puerperal fever was spread from patient to patient by doctors. Yet in spite of the compelling evidence Semmelweis met with strong opposition. Eventually, in 1850 he left Vienna for the university hospital in Pest where, as professor of obstetrics he was responsible for the care of mothers during childbirth. Again he enforced what are now recognised as antiseptic practices and the number of women dying from puerperal fever after having a baby in Pest fell to 0.8%. But yet again his findings and publications were resisted, not just in Hungary but also abroad.

Why was there opposition to Semmelweis?

Pain and suffering during childbirth was an accepted part of European culture in the 18th and 19th century (Figure 1.3). It was hard for doctors to admit that they themselves had spread the disease and killed their patients instead of curing them. To change their point of view would mean accepting that the deadly disease was caused by a transferable agent.

Figure 1.3

An engraving by Holbein showing an angel, accompanied by Death, chasing Adam and Eve from the garden of Eden. In Genesis, the first book of the Bible, God says to Eve, "I will increase your trouble in pregnancy and your pain in giving birth". Some doctors preferred to accept this as an explanation of childbed fever rather than Ignaz Semmelweis's more scientific explanation, which he demonstrated clearly in two different hospitals.

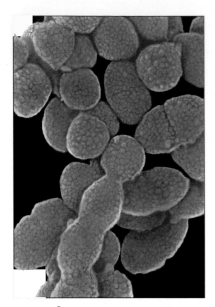

Figure 1.4

By the mid-20th century it was shown that puerperal fever was caused by the Streptococcus pyrogenes bacterium and that antibiotic drugs could destroy it. In developed countries of the world the fear of death from childbed fever was finally lifted from women and their families. This picture shows cells of the bacterium dividing. This strain is resistant to common antibiotics. The magnification is about 21 500 times life size.

Questions

8 What part might the reputation and personality of doctors have played in delaying the acceptance of Semmelweis's ideas?

9 Why did the disease linger in the UK, Europe and the USA until the middle of the 20th century?

10 Why do thousands of women in developing countries of the world still die each year from puerperal fever? Is this acceptable and how could these deaths be prevented?

Another factor might have been that handwashing probably seemed rather an odd practice at the time. There was no indoor plumbing, so getting water to wash in was not easy. Water brought in would have been cold, and the chemicals used to wash with (such as chlorinated lime) would have eventually damaged the skin of the hands. It is difficult to imagine from the perspective of the 21st century just how difficult such a simple procedure must have seemed in the 19th century. Even outside of hospitals, women died of childbed fever. However, as the American Oliver Holmes pointed out, only the patients of certain doctors died. So if the doctor delivering a baby was clean, washed his hands and changed his clothes regularly, then the mother would probably live. If not, she had a high chance of developing childbed fever.

Semmelweis found the rejection of his work unbearable, because he recognised that simple hygiene measures held the key to saving thousands of lives. By the 1860s he suffered a major breakdown and went to a mental asylum in Vienna. In 1868, aged only 47, he died – by an ironic twist of fate from an infection picked up from a patient during an operation.

A lingering threat

Relying on handwashing alone leaves a great deal of room for human error. Dying of fever after childbirth remained a feared outcome until after the Second World War, when antibiotics became widely available (Figure 1.4). However, in some parts of the world even today, where knowledge of micro-organisms is scanty and antibiotics are rarely available, mothers are still dying of puerperal fever in the days immediately after they have given birth.

Cholera in London

A new disease

A number of severe cholera outbreaks gripped London and other parts of the country in the 1830s and 1840s, Semmelweis and others were trying to convince people that childbed fever was infectious.

Cholera was a new disease in Britain. People with cholera suffer severe stomach pains; they produce vast amounts of very watery diarrhoea and sometimes vomit as well. The disease spreads very quickly through a community. The cholera epidemic of 1831–2 killed 32 000 people in 3 months

The people who suffered most from cholera were the poor in large towns, who lived in overcrowded homes with poor sanitation. At the time there was no system for removing household rubbish other than throwing it into the streets.

Most doctors then thought that diseases such as cholera were spread either by touch or by 'bad air' (miasma). One campaigner, Edwin Chadwick, used the miasma theory as the basis for a campaign to clean the rubbish from the streets and to build sewers (Figure 1.5).

In this period John Snow was working as a doctor, first in Newcastle where he saw the terrible effects cholera on miners in the 1831

Figure 1.5

The campaign to build sewers led to major engineering projects in London This engraving shows the main northern sewer under construction in 1859. A proqramme to build 156 km of sewers was finally completed in 1865.

epidemic, and then in London. He became very interested in what was happening and made careful observations of all the cholera cases he came across. Snow traced one outbreak of the disease to the arrival of one sick seaman from Hamburg.

Snow came to the conclusion that cholera was actually caused by a poison which reproduced in the human body and was found in the vomit and diarrhoea of cholera patients. His hypothesis, as he stated it, was that 'disease is communicated by something that acts directly on the alimentary canal. The excretions of the sick at once suggest themselves as containing some material which being accidentally swallowed might ... multiply itself'.

In 1849 Snow published a pamphlet explaining his theory that the main way in which the disease was passed on was through water contaminated with an infectious agent. However he was not the only person trying to come up with an explanation for the killer disease at the time. At this stage most people did not believe him and his pamphlet was largely ignored.

The Broad Street pump

In 1854 another major cholera outbreak hit London. Snow undertook a detailed record of all the cholera cases he could find. His meticulous documentation showed up a number of different and important features.

In one small neighbourhood under John Snow's surveillance the number of cholera deaths was terrifying. Within a little over 200 metres of the junction of Cambridge Street and Broad Street, 500 men, women and children lost their lives in 10 days. By plotting the homes of all the cases on a local map he saw that they all got their drinking water from

Questions

11 Which aspects of the cholera epidemic were explained by the 'bad air' theory?

12 Suggest reasons why the authorities burnt barrels of tar in the streets and issued lime to whitewash houses during epidemics?

13 Edwin Chadwick argued that sewers should be regularly flushed with water to get rid of the smells. In London this washed the sewage into the Thames. Why did he believe that this would improve public health? In fact it may have increased the incidence of cholera. Can you explain why?

14 Snow introduced the idea that the infectious agent might be able to multiply itself. Why was this idea an important feature of his theory?

15 In the 1830s, microscopes were powerful enough to see tiny organisms in water but they were not good enough to detect germs. What influence do you think this had on the early rejection of Snow's theory?

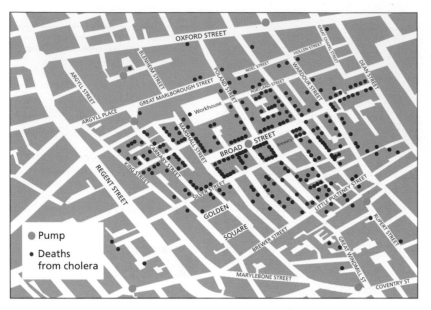

Figure 1.6

When John Snow plotted all the cholera deaths in a neighbourhood he found that they clustered around one water source – the Broad Street pump.

Pump
Deaths from cholera

the same source – the Broad Street pump (Figure 1.6). For Snow this confirmed his ideas. He persuaded the panic-stricken local officials to have the handle of the pump removed. Once this was done the epidemic was contained and began to subside.

Snow's records showed him something else as well. In the same year he had studied a region of South London where houses in the same streets were supplied by two separate water companies. The Lambeth Water Company took water from the upper reaches of the Thames, before it flowed through the city. The Vauxhall Water Company took water from the lower Thames, after it had passed through London and received most of the city's sewage. People supplied with water by the Vauxhall company were far more likely to develop cholera. As John Snow wrote of the Vauxhall supply in the *Medical Times* of 1854: 'Part of the water has passed through the kidneys and bowels of 2 million and a quarter of the inhabitants of London.'

As a result of John Snow's evidence people began to take his ideas seriously. We now know that his description of the disease was very near the truth. Finally, in 1883 in a cholera epidemic in Cairo, Robert Koch identified the cholera bacterium in the victims of the disease, in water and in food.

Pasteur and microbes

The work of Semmelweis and Snow led gradually to the acceptance of the idea that infectious diseases are brought about by an infectious agent (or 'germ') which is transferred from one individual to another and causes disease. But entrenched views of the causes of disease were hard to shift. It took many years of work by many people working in many countries before there was widespread acceptance of the theory. Two people who are now remembered for their part in establishing the theory were the French scientist, Louis Pasteur and the German doctor, Robert Koch.

Pasteur was born in 1822 and lived until the age of 72. Pasteur was always ambitious to make scientific discoveries. As a professor of chemistry at Lille he became interested in fermentation because of problems encountered by local vinegar makers. There was a new theory that fermentation was caused by microscopic yeast cells and was not simply a chemical reaction. Pasteur studied fermentation in great detail and produced the evidence to persuade people that the yeast theory was correct. As a result he became very interested in microbes and where they come from.

Questions

16 Summarise John Snow's theory and show how it differed from the other explanations for the spread of cholera.

17 Why was the removal of the handle of the Broad Street pump so effective in persuading people that Snow's theory was valid?

18 Why did Snow's studies of water supplies in South London support his theory but not the miasma theory?

19 In what way was Snow's explanation for the spread of cholera incomplete when he published his theory?

20 What were the similarities and differences between the way Semmelweis developed an understanding of the cause of childbed fever and the way Snow worked out what was happening in the spread of cholera?

21 Cholera and other diseases which cause diarrhoea are still the main killers of children throughout the world. Why do people die of cholera, and why do you think children are particularly vulnerable?

Figure 1.7

Pasteur dictating notes to his wife. Louis Pasteur was a major figure in the development of the germ theory of disease. His wife recorded his dictation, wrote up his notes and discussed his ideas with him.

Key terms

Microbes (or micro-organisms) are minute living beings which are only visible with the help of a microscope. We now know that yeasts are the microbes which cause fermentation and bacteria are the microbes which cause diseases such as childbed fever and cholera.

A **germ** is a microbe which can cause disease.

At the time there was a widespread belief that living things could arise from dead things. This was the theory of spontaneous generation. A number of scientists had studied the problem but there was no consensus; despite the work of the Italian biologist Lazzaro Spallanzani who had carried out a series of experiments in 1768 which he claimed showed that microbes must develop from other microbes and not by spontaneous generation.

In 1859 Pasteur decided to join the debate. He realised that it would be difficult to prove that spontaneous generation never happened but he intended to show that there was no evidence that it did happen. This led to his series of classic experiments with swan-necked flasks showing that that the micro-organisms which grew in broth, turning it cloudy and mouldy, did not appear by spontaneous generation but were already present in the air (Figure 1.7).

In 1866 Pasteur showed that microbes could be the cause of disease. At the time a killer disease was destroying large numbers of silkworms in the French silk industry. Pasteur was able to extract microbes from the bodies of dead silkmoths and show that they were the cause of the deaths. He recommended a system for culturing eggs only from healthy moths. By keeping the healthy eggs away from all contact with living caterpillars he made sure the eggs would produce moths free of disease (Figure 1.8).

Identifying germs

Pasteur had established the germ theory of disease but he did not know how to identify the different kinds of germ. In 1866, when Pasteur was already in late middle age, the young German doctor Robert Koch became interested in examining microbes with the more powerful microscopes which were becoming available (Figure 1.9). He devised the techniques for culturing bacteria on agar jelly. He also worked out how to study the bacteria using dyes to stain them on glass slides so that they could be seen and recognised under a microscope. With these methods, he and his fellow workers discovered the causes of eleven diseases including anthrax (in 1863), tuberculosis (TB) (in 1882) and cholera (in 1883) (Figure 1.10).

Figure 1.8

Drawing of a healthy silkworm from Pasteur's book Diseases of Silkworms.

Question

22 Pasteur believed that spontaneous generation never happens. Why is it impossible to 'prove' that something never happens (to prove a negative)? What did Pasteur set out to show instead?

Figure 1.9

Robert Koch in his laboratory.

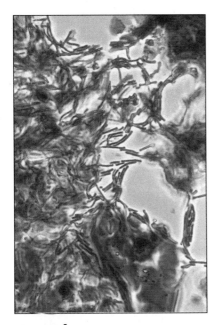

Figure 1.10

Photograph taken through a light microscope showing tissue infected with the bacterium Bacillus anthracis. The dye added to the specimen stains the bacterial cells blue. Robert Koch was the first person to isolate this bacterium which causes anthrax.

Question

23 Explain why the development of more powerful microscopes led to the final acceptance of the germ theory of disease.

Koch observed the germs multiplying and investigated the conditions which would stop them reproducing. In many ways Koch was the father of the science of bacteriology and in 1910 he was awarded the Nobel Prize for medicine. By the end of the 19th century the idea that infectious diseases were caused by germs was almost universally accepted.

Germs today

As a result of the work of scientists such as Semmelweis, Snow, Pasteur and Koch, a picture of the nature of infectious diseases began to be developed, a model which has moved much further forward though the 20th century, aided by ever more sophisticated technology.

Infectious diseases are widespread all over the world and affect not only humans but also all other animals and plants. They range from minor inconveniences such as the common cold and mild sickness through to devastating and fatal illnesses such as AIDS, yellow fever and malaria. People catch these diseases when an infectious agent (or germ) invades their bodies. Germs cause tissue damage as they reproduce themselves in or on the body, and it is this which gives the symptoms of disease.

Cells as the basic units of living things

Figure 1.11

All living organisms are made up of cells, discovered in the very early days of the microscope. These cells carry out all the basic functions of life – making new proteins and materials for growth, transferring energy from food in respiration, replicating to form new cells, getting rid of waste material and so on.

The mechanisms by which these processes take place are similar in all living organisms.

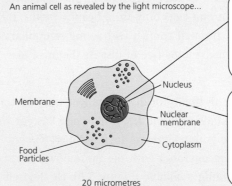

An animal cell as revealed by the light microscope…

All cells have features in common, and ways in which they differ. Animal cells have a nucleus which contains the genetic material (in the form of DNA), which controls how the cell works.

Chemicals can enter and leave the cell through the membrane. Incoming chemicals are the ones the cell needs for respiration and growth including oxygen and glucose. Outgoing chemicals are waste products such as carbon dioxide.

Membrane

Nucleus

Nuclear membrane

Cytoplasm

Food Particles

20 micrometres

Bacteria and viruses

As microscopes improved Pasteur, Koch and others could see bacteria, but it wasn't until the electron microscope was developed that scientists could identify viruses.

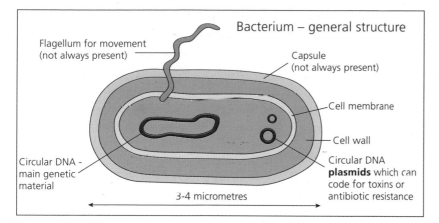

Bacterium – general structure

Flagellum for movement
(not always present)

Capsule
(not always present)

Cell membrane

Cell wall

Circular DNA -
main genetic
material

Circular DNA
plasmids which can
code for toxins or
antibiotic resistance

3-4 micrometres

Figure 1.12

Tiny they may be, but some bacteria can wreak havoc on the human body and cause death in a matter of hours.

Question

24 In a disease caused by bacteria the symptoms tend to come on gradually, getting steadily worse. In diseases caused by viruses the symptoms come and go – for example, the temperature goes right up, then falls a bit, then shoots up again. Suggest an explanation for this difference.

Bacteria are single-celled organisms (Figure 1.12). Most bacteria are either harmless or beneficial, but some invade human tissues and grow to form colonies in certain organs causing disease. Examples of bacterial diseases include tonsilitis, TB and cholera.

In contrast, viruses are not really independent organisms but 'packets' of DNA with some enzymes contained in an outer case of protein (Figure 1.13). Viruses cannot survive on their own and they can only multiply by invading a healthy cell. Once inside the host cell the virus takes over the cell's biochemistry and uses it to produce new copies of itself (Figure 1.14). This process continues until the host cell structure completely breaks down, releasing the new viruses to infect other cells.

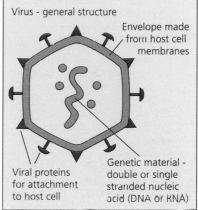

Virus - general structure

Envelope made from host cell membranes

Viral proteins for attachment to host cell

Genetic material - double or single stranded nucleic acid (DNA or RNA)

Figure 1.13

Even tinier than bacteria, viruses nevertheless cause some of the deadliest diseases to affect the human race.

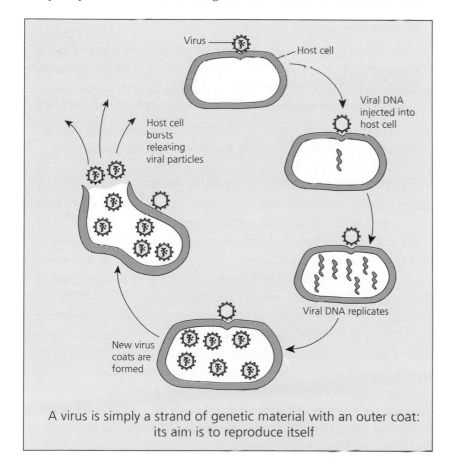

Virus

Host cell

Viral DNA injected into host cell

Host cell bursts releasing viral particles

Viral DNA replicates

New virus coats are formed

A virus is simply a strand of genetic material with an outer coat: its aim is to reproduce itself

Figure 1.14

A virus multiplying inside an infected cell.

25 Use the information about the way diseases are spread to explain the observations of Semmelweis and Snow concerning childbed fever and cholera.

26 In Snow's time some people thought cholera and other infectious diseases were spread by bad air. Others thought that diseases were spread by touch. Snow's theory was that the infection was carried by something in contaminated water. How does our modern understanding of the spread of infectious diseases help to explain why these earlier theories might have seemed to provide explanations at the time? ·

27 For each way of spreading disease, decide how you might set about showing that there is a causal link between the method of transmission and people getting a disease. How, for example, could you show that people catch skin infections from infected clothing?

It is this cell destruction and the body's reaction to it which gives rise to the symptoms of viral diseases such as polio, influenza and AIDS.

The cells of bacteria and the structure of viruses are quite distinctive. Often a doctor can diagnose a disease simply from the symptoms, but sometimes pathologists have to culture the infected tissue to identify the particular infectious agent involved. Because bacteria and viruses are so different it is relatively easy to tell them apart with microscopes.

How do diseases spread?

Once the germ theory of disease was accepted people could really begin to understand how infectious diseases are spread from one person to another. There are all sorts of ways in which germs can be passed on – and knowing this can help people avoid spreading disease.

- Living creatures can transmit infection from one person to another. Mosquitoes, for example, spread malaria.

- Inanimate objects like clothing and bedding can carry germs from one person to another. For example, *Staphylococcus* infections are often caught from hospital bedding.

- Direct contact is often important in the spreading of skin diseases and sexual diseases such as impetigo and syphilis.

- Whenever someone coughs, sneezes or talks millions of germ-containing droplets are expelled from the respiratory tract to be inhaled by someone else. Diseases such as influenza, measles and tuberculosis are spread like this.

- Many of the pathogens which cause gut diseases are transmitted by contaminated food or drink. Germs from faeces, urine and vomit can also be spread if people put their dirty hands in their mouths. Examples include most forms of diarrhoea, including cholera.

- Germs can also enter the body directly through cuts in the skin or wounds – for example, hepatitis B, AIDS, rabies and tetanus.

A 20th-century controversy over germs

All of the early scientists who tried to show that the cause of infectious disease was germs struggled to get their views accepted. It might seem likely that with hindsight lessons would have been learned. But as recently as the 1980s a similar problem appeared again. For many years, doctors thought that gastric and duodenal ulcers were lifestyle diseases. The theory was that stress caused excessive excretion of acid in the stomach. This in turn caused an ulcer, a very painful area where the lining of the gut erodes. Treatment was with drugs to reduce the amount of acid produced or to neutralise it, but when the treatment stopped the ulcers returned and eventually surgery was the only option.

In the 1980s an Australian, Dr Barry Marshall, found the bacterium *Helicobacter pylori* living in the stomachs of all his patients with gastric ulcers. The scientifically accepted view was that bacteria could not live in the acidic conditions of the stomach, so this finding surprised Marshall.

He suspected that the bacteria might be causing the stomach ulcers and devised a series of tests to demonstrate whether this was true. Eventually he developed a treatment based on using existing cheap antibiotics which cleared the ulcer symptoms in his patients, and because it destroyed the bacteria the ulcers did not return. Other members of a family could be checked for the bacterium, which was passed from one to another, and treated before symptoms developed.

The idea that the majority of ulcers were yet another example of the germ theory of disease with a cheap and simple treatment met with a great deal of resistance among other doctors, particularly those who spent a great deal of their time treating long-term ulcer sufferers, and from drug companies. Drugs to relieve ulcer pain and block acid production have been among the best-selling medicines in the world. It took 10 years to convince about 10% of doctors that Marshall was right – but now the great majority of physicians accept that *H. pylori* rather than stress is the major cause of ulcers. Barry Marshall went on to show that there is a link between the same bacteria and certain stomach cancers, but although bacteria had never before been linked to cancer, *H. pylori* has now been classified as a class 1 cancer causing agent (carcinogen).

Questions

28 Barry Marshall suspected that the bacteria he found in the stomachs of his ulcer patients might be causing the ulcers. How would you set about proving such a hypothesis?

29 There was enormous resistance to Marshall's ideas at first. Why do you think there was such strong opposition from doctors who specialised in the treatment of ulcer patients?

30 It took more than 10 years for Marshall's ideas about the role of bacteria in the formation of ulcers to be generally accepted. However, his equally ground-breaking work on the role of the same bacteria in some stomach cancers was accepted much more readily. Why do you think this was the case?

31 Find examples from the case studies in this chapter to illustrate the five key terms in Figure 1.15 on page 12: data, evidence, patterns, generalisations and explanations.

Review Questions

32 Draw up a timeline showing how ideas about the cause of diseases developed during the 18th, 19th and 20th centuries.

33 People believe in the 'germ theory of disease' in a way in which they probably do not accept other theories. People clean toilets and don't like to drink out of dirty mugs or glasses.

- Why do people accept germ theory so completely?

- In spite of believing in germ theory, people are often very sloppy when it comes to avoiding food poisoning: buying chilled, cooked foods, eating food which contains raw eggs, etc. Why doesn't our belief in the germ theory motivate us to take all the recommended measures to avoid food poisoning?

34 A Bavarian doctor who did not believe in the germ theory of disease deliberately consumed a culture containing millions of cholera bacteria. A Russian pathologist did the same thing at about the same time. So did several other people. None of these people developed cholera. Why are these experiments seldom mentioned? How could someone who believes in the germ theory of disease account for these results?

Discussion point

The work of Semmelweis, Pasteur and Marshall all show that it is not simply the validity of the science which decides whether an idea is accepted. Factors such as the reputation of the scientist, any clashes with accepted social standards, financial and other vested interests and who has the loudest voice or the best connections all play a part. Is this inevitable? Argue your case either for or against, and make suggestions for ways in which the situation might be changed.

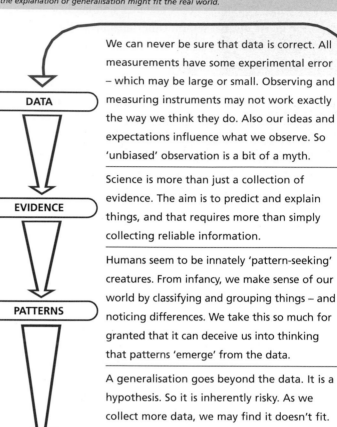

Before you can begin to understand anything you need to know about it. You need data. Data are observations and measurements. People can usually reach agreement on the data, even if they dispute what it means. If anything in science can be called 'facts', then it is the data.

Data is not automatically evidence. Scientists need to decide if it is reliable and relevant to the question in which they are interested. Only then does it become evidence.

Patterns mean things like similarities, differences, trends, relationships. Patterns in data may indicate something worth looking into further.

If scientists spot a pattern, they may speculate that it would apply to other cases, as well as the ones they have already looked at. That is, they generalise beyond the cases they have studied. If they propose that a pattern applies to all cases, it is a universal generalisation. Generalisations are useful because they allow us to predict what will happen in other, yet unexplored, cases.

Science is about finding explanations. These take the form of theories, often based on models of what is thought to be going on. Finding an explanation involves imagination and creativity. Although it is important to have data as a basis for an explanation, it doesn't 'emerge' automatically from the data. An explanation is a hypothesis, or conjecture, about what is causing things to happen as they do. Useful theories are ones which make precise and testable predictions. If these are found to be correct, it increases confidence in the explanation. If not, the explanation may have to be revised.

We can never be sure that data is correct. All measurements have some experimental error – which may be large or small. Observing and measuring instruments may not work exactly the way we think they do. Also our ideas and expectations influence what we observe. So 'unbiased' observation is a bit of a myth.

Science is more than just a collection of evidence. The aim is to predict and explain things, and that requires more than simply collecting reliable information.

Humans seem to be innately 'pattern-seeking' creatures. From infancy, we make sense of our world by classifying and grouping things – and noticing differences. We take this so much for granted that it can deceive us into thinking that patterns 'emerge' from the data.

A generalisation goes beyond the data. It is a hypothesis. So it is inherently risky. As we collect more data, we may find it doesn't fit. As with pattern-seeking, though, this seems to be a way that humans think. We don't have to put our hands into very many fires to conclude that they all burn! But there is no way of reasoning that ensures that any generalisations we arrive at are correct.

In practice it is often difficult to know what to do if data do not agree with a prediction. It might be due to experimental error in the data, or to some other factor that hasn't been taken into account. So it might be wiser not to reject an explanation too quickly, if it has other advantages – especially if there is no other alternative explanation!

Chapter 2

Preventing diseases

The issues

Governments have to consider how best to deploy the resources available for health care. The answers differ from one country to the next. Primary health care and programmes to prevent disease can be more cost effective than building hospitals and setting up expensive specialist services, but political pressure from those with money and influence in society may mean that the interests of the better off dominate over the needs of the majority.

The science behind the issues

If we know which micro-organisms cause a disease, and where these can be found, then the best way of preventing disease is to eliminate the causes of infection. Vaccination is also a very effective way of preventing disease because it stimulates the immune system to produce antibodies which protect against future infection. Vaccination, however, has its limits against diseases such as influenza because new forms of the virus keep appearing and each variety needs its own vaccine.

What this tells us about science and society

The discovery of vaccination shows how both observation and experiment are important in testing ideas. Explanations do not 'emerge' from the data automatically – but are arrived at through imagination and conjecture. Scientists are often reluctant to give up an established explanation, which has shaped their work and their thinking for many years, even when evidence starts to mount up against it. Since explanations cannot be 'deduced' from the data, but are conjectures which account for it, there is always room for disagreement about what the evidence really implies.

Decisions about whether or not to apply scientific knowledge can highlight the tension between the rights of individuals and the interests of society as a whole (Figure 2.1).

Figure 2.1

A child being vaccinated in Guatemala. Vaccination programmes to prevent diseases are only fully effective if nearly all children are immunised. Yet there are parents who feel strongly that the risks of vaccination are too high. If too many parents ignore the advice of doctors they put at risk the health of many children because they increase the chance of an epidemic. Science alone cannot resolve such issues.

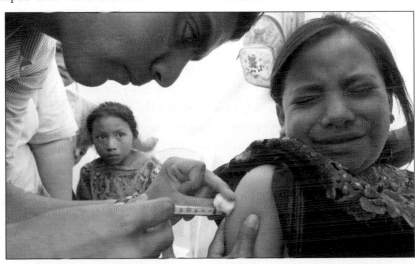

Figure 2.2

Some of the barriers to prevent infection.

Resistance to infection

Doctors and medical science have played their part in the conquest of disease but even now many diseases cannot be cured. Doctors are often only able to speed recovery or relieve discomfort.

In John Snow's time, most deaths, especially deaths in childhood, were caused by infectious diseases such as typhoid, diarrhoea, whooping cough, measles, scarlet fever and TB. These diseases flourished in big cities where the children of the poor were underfed, lived in crowded homes with poor sanitation and drank polluted drinking water.

Ref to pages 4–6

At first the remedies arose from a better understanding of the germ theory of disease leading to policies to limit infection. Success depended on social action by politicians and others to relieve poverty and improve housing. Sanitary engineers played their part by installing piped water supplies and constructing enclosed sewers. Nutritionists played their part by research which helped to develop guidelines for the diet of children. All this activity greatly cut down the deaths from infections long before vaccination and antibiotics were available.

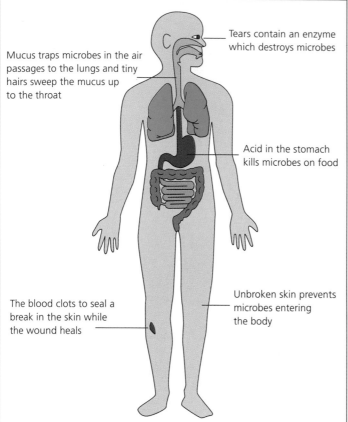

Tears contain an enzyme which destroys microbes

Mucus traps microbes in the air passages to the lungs and tiny hairs sweep the mucus up to the throat

Acid in the stomach kills microbes on food

The blood clots to seal a break in the skin while the wound heals

Unbroken skin prevents microbes entering the body

When people are healthy they have considerable resistance to disease. As Figure 2.2 shows, the human body can defend itself against infection.

People do not necessarily fall ill even if bacteria or viruses manage to break through and invade their bodies. The white cells in the blood provide the next line of defence. Some white cells attack and destroy the germs while others produce antibodies which can bind to the invading micro-organisms and kill them (Figure 2.3).

Antibodies not only help to fight off infection, but they may also stay in the body for months or years. This means that, after infection, people become relatively immune to further infection by the same disease. Also the immune system becomes more responsive, so that it effectively learns to make antibodies against the particular germ and can respond much more quickly to future infections.

These natural defences against infectious diseases are much weaker if people are suffering from malnutrition. Diarrhoea and respiratory illnesses can kill undernourished children but seldom children who are well fed. Measles is another infection which can become a killer disease for people of all ages, if they are starving. A good diet is per-haps the best protection against the infectious diseases which can threaten life.

Key terms

The white blood cells, and the antibodies which some of them produce, are part of the body's **immune system** to fight infection.

White cells produce **antibodies** which are specific; each type of antibody targets a particular bacterium or virus.

The defence mechanisms of the immune system can be stimulated artificially by **immunisation**. This is **vaccination**. A **vaccine** gives immunity to a particular infection.

Question

1 Why do the body's defences sometimes fail (see Figure 2.2)? Give examples.

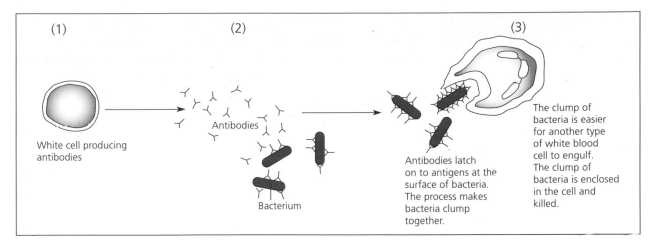

(1)   (2)   (3)

White cell producing antibodies

Antibodies

Bacterium

Antibodies latch on to antigens at the surface of bacteria. The process makes bacteria clump together.

The clump of bacteria is easier for another type of white blood cell to engulf. The clump of bacteria is enclosed in the cell and killed.

Figure 2.3

How white cells destroy invading bacteria.

Tuberculosis

The impact of improved living conditions and better diet is illustrated by the disease tuberculosis (TB). TB has always been a very common human infection. The disease still kills over 2 million people each year around the world.

TB comes in many forms, but it is most commonly caused by infections of *Mycobacterium tuberculosis*. When infectious people cough, sneeze, talk or spit they spread TB germs into the air. An untreated person with active TB is likely to infect 10 to 15 other people each year. Crowded living or working conditions add to the ease with which it spreads from person to person. Infection, however, does not necessarily lead to sickness with the disease. The immune system gives protection and the bacteria may lie dormant in the body protected by a thick waxy coat.

The World Health Organisation (WHO) estimates that overall one-third of the world's population is currently infected with TB, but that only 5–10 per cent of infected people fall sick or become infectious at some time during their life.

The other common source of infection is from *Mycobacterium bovis*. This bacterium affects cattle, and people become infected and develop TB from drinking infected milk. It is only a problem where the disease has not been controlled in cattle and where milk is drunk unpasteurised.

TB can affect many areas of the body, including the lungs and the bones, but the most common forms of TB affect the respiratory system. The bacteria stimulate the action of the immune system, making the body damage and destroy its own lung tissue (Figure 2.4).

Typical symptoms of TB are fever, night sweats, the inability to eat and loss of weight. In response to the damage in the lungs there will often be a cough which will produce mucus from which *M. tuberculosis* can be cultured. In severe cases the mucus is blood-stained.

In many developed countries the number of cases of TB has fallen dramatically over the last 150 years.

Improving living standards is the most effective way of controlling TB. Less crowded housing and working conditions mean people are less

Key term

Almost all milk drunk in the UK is **pasteurised**. Heating the milk for a short time destroys micro-organisms such as the bacterium which causes TB. The process is named after Pasteur who suggested the process to French wine-makers to stop their wine going bad while it was maturing in casks.

Figure 2.4

Coloured X-ray of the lungs of someone infected with TB shown by the fluffy yellow areas in the patient's left lung.

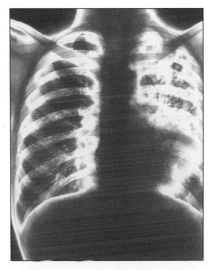

2 Study Figure 2.5.

a Which of the lines of the graph would you expect to be based on more reliable data and why?

b Which important methods of controlling the spread of the disease are not indicated on the graph?

c How do you account for the decline in the incidence of TB from 1913 to 1940?

d Suggest reasons for the rise in the number of notifications of the disease between 1940 and 1950.

e What role did mass X-ray screening of the population play in helping to lower the incidence of the disease? Why has the mass X-ray programme ended in the UK?

f Has prevention or cure played the greater part in the reduction in the notification?

3 Give examples of ways by which improved social conditions can help to reduce the spread of TB. Explain why the changes are effective.

4 Why is TB no longer common in countries such as the UK but is still widespread in other parts of the world?

Figure 2.6

A photograph taken in 1973 showing a young Bangladeshi with smallpox in a relief camp. Smallpox was caused by the varola virus. The patient's body is covered by large spots. Survivors of the disease were usually scarred for life.

likely to pass on the disease. Generally healthier and better-fed people are less likely to develop debilitating TB, even if they meet *M. tuberculosis*. Preventing and treating the disease in cattle along with pasteurising or heating milk before it is drunk both prevent the spread of *M. bovis*. As Figure 2.5 shows, vaccination also has a part to play in reducing the numbers suffering from the disease.

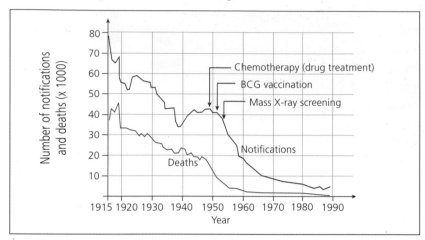

Figure 2.5

The incidence of respiratory tuberculosis in England and Wales from 1915 to 1990.

Immunisation

Immunisation helps to prevent sickness by using the body's own immune system to fight infection. The first disease to be tackled in this way was smallpox. Many people died if they caught smallpox and even if they survived they were disfigured with pox scars (Figure 2.6). It has been known for hundreds of years that once someone has had smallpox they could not catch the disease again. This led to experiments to explore the possibility of immunising people with a mild attack of the disease. As long ago as 1718, Lady Mary Wortley Montagu came across this approach to prevention in Turkey and returned to England to try out the idea. The first people tested were criminals from Newgate prison under sentence of death. They were offered their freedom if they agreed to be immunised with smallpox. All survived and were set free. But it turned out that the technique worked with some people but not with others who developed a serious attack of the disease and died or infected others. As a result this method of immunisation was made illegal in England by 1840.

Jenner and cowpox

The story of banishing smallpox began in the county of Gloucestershire, England, where the country doctor Edward Jenner (1749–1823) practised medicine. His ideas were based on his everyday experiences of visiting the sick among the farming community.

The local farmers probably set Jenner's mind to work on the problem of smallpox. They told him that the girls who milked the cows often suffered from the mild disease cowpox but rarely caught the more severe and often fatal smallpox. Cowpox comes from handling cows and causes spots, especially on the hands.

Spots are also a symptom of smallpox, and Jenner wondered if deliberately infecting people with cowpox, would protect them from smallpox. Jenner did not test his theory until 1794 which was a year in which one in five of all reported deaths were the result of smallpox.

He took pus from cowpox spots on the hands of Sarah Nelmes and, with a needle, scratched it into the arm of James Phipps, a healthy boy (Figure 2.7). The boy recovered from the cowpox and two months later Jenner went one risky step further and scratched pus from the spots of a smallpox victim into the boy's arm. Fortunately, the boy did not develop smallpox. His survival helped confirm the idea that exposing a person to a mild dose of a disease stimulated them to resist its more serious forms. Jenner had shown that inoculating someone with a mild disease could give protection against a fatal illness.

Then as now, vaccination raised people's fears (Figure 2.8). Smallpox vaccination was not just the beginning of the story. The eradication of smallpox has also been the greatest triumph so far for vaccination programmes. As recently as the 1960s, smallpox killed 2 million people, and infected 10-15 million more, disfiguring many of those who did not die. Mostly through the efforts of the World Health Organisation (WHO), £20 million ($33 million) were spent in a world-wide vaccination programme between 1967 and 1977. Previously it cost much more to treat people and impose quarantine restriction to help prevent the disease from infecting even more people. In May 1980 the WHO was able to declare that the world was free of the scourge of smallpox.

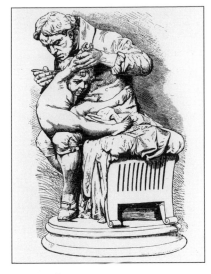

Figure 2.7

Drawing of a statue of Edward Jenner showing him infecting James Phipps with cowpox.

Key term

The term **vaccination** is derived from the Latin for cow. It reminds us of the importance of cowpox in giving medicine a powerful weapon to prevent disease.

Questions

5 What fears are illustrated by Figure 2.8? How do they compare with the worries people have about vaccination today?

6 What was Jenner's theory? How did he test his theory? What part did prediction and experiment play in his discovery?

Figure 2.8

This Gillray cartoon illustrated people's nightmares about vaccination suggesting appalling results.

Pasteur and chicken cholera

Following his successful investigations of fermentation and the disease of silkworms, Pasteur had established a substantial reputation for solving practical problems and farmers sought his advice. He knew about Jenner's work and had wondered whether or not vaccination could be used to prevent other diseases. *Ref to pages 6–7*

In 1879 Pasteur was studying a disease called chicken cholera. He had identified the microbe which causes the disease and succeeded in culturing it in the laboratory. He and his assistants were injecting healthy chickens with the culture and showing that microbes grown outside the body did infect healthy chickens so that they rapidly became ill and died.

One day, by mistake, an assistant injected a batch of chickens with a culture of microbes which had been left standing for several weeks during a holiday. The chickens were ill for a time but then recovered. The assistant was about to ignore this experiment. Pasteur, however, told him to give the chickens a second injection – this time using a fresh culture of the microbes. The birds stayed healthy.

Further experiments confirmed Pasteur's idea that the microbes in the culture which had stood over the holiday had become weakened. They were only able to give chickens a mild form of the disease. Once the chickens had recovered from the mild infection they became immune to the fresh injection. Pasteur had discovered the basis of vaccination using a weakened (attenuated) form of the infective agent.

Pasteur and anthrax

Following his work on chicken cholera, Pasteur had set out to develop a vaccine for anthrax. Anthrax is a disease of animals which is highly infectious. In the 1870s up to 50% of all the sheep and cattle in France had been dying of the disease. Animals have only to graze over the burial site of anthrax victims to have a strong chance of catching the disease. Pasteur showed that an infectious agent was being brought up from the buried corpses by worms.

Pasteur enjoyed publicity and was willing to make enemies in the medical profession. He enjoyed convincing his opponents that they were wrong. His quest for a vaccine for anthrax gave him a famous opportunity to do so. By this stage in his life Louis Pasteur had suffered a very severe stroke. Although he recovered and his mind was unaffected, his speech, gait and ability to use his hands was never the same again – for the rest of his life Pasteur relied heavily on his team of trusted fellow scientists to carry out the experiments he dreamed up.

Pasteur and his team were confident that they could find a way of beating anthrax – but it proved more difficult than they thought. The anthrax germ proved very hard to grow in the laboratory.

Meanwhile Robert Koch had developed a way of culturing anthrax spores. Pasteur immediately used this new technique to grow anthrax bacteria and then tried to make a vaccine. None of the methods he attempted seemed to give reliable results.

Question

7 Pasteur once famously said that: 'In the field of experimentation, chance favours only the prepared mind.' How is this is illustrated by his work with chicken cholera?

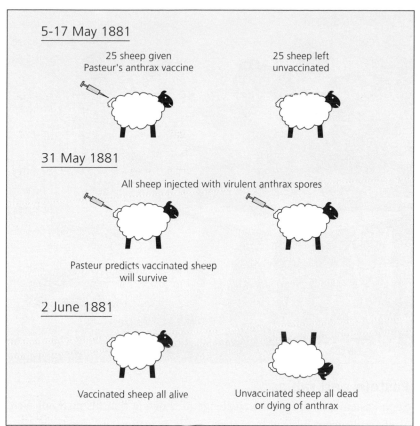

5-17 May 1881

25 sheep given
Pasteur's anthrax vaccine

25 sheep left
unvaccinated

31 May 1881

All sheep injected with virulent anthrax spores

Pasteur predicts vaccinated sheep
will survive

2 June 1881

Vaccinated sheep all alive

Unvaccinated sheep all dead
or dying of anthrax

Figure 2.9

Pasteur's success at Pouilly-le-Fort silenced almost all critics of his germ theory – the evidence was there for all to see.

Then Toussaint, a young vet, claimed to have produced a successful vaccine using a different method. Pasteur claimed that it was unreliable, and quickly announced his own vaccine was on the way (produced in a very similar way to that described by Toussaint). What Pasteur and his assistants did was to keep anthrax bacteria warm at just over 40°C. After 8 days the microbes were much weakened and no longer able to cause the fatal disease. They could, however, be used for vaccination.

As a senior and highly respected scientist, Pasteur's version was accepted and Toussaint retired, a broken man. But Pasteur's steady work on anthrax was then interrupted. Hippolyte Rossignol, a vet who had little time for the germ theory of disease, threw down a very public challenge to Pasteur. Rather than appear unsure of his work, Pasteur accepted the challenge, although he was not at all certain his vaccine would work.

The trial took place on Monsieur Rossignol's farm at Pouilly-le-Fort (see Figures 2.9 and 2.10). A crowd gathered to watch the start of the experiment. The farmers, vets and doctors returned on 2 June and applauded Pasteur as he arrived to find all the vaccinated sheep alive and well while all the unvaccinated sheep had died.

After the success of the trial at Pouilly-le-Fort, Pasteur completed the development of his vaccine against anthrax. The vaccine had a major effect on farming for generations to come – but the confirmation of the germ theory of disease was to have even greater implications for the health and well being of people all over the world.

Discussion point

Pasteur's version of an anthrax vaccine was accepted rather than the one proposed by the young vet Toussaint. Why do you think this was the case? What are the dangers of considering the personal characteristics of scientists when we decide whether or not to accept their ideas? *Ref to pages 18–19*

Question

8 Why was the trial at Pouilly-le-Fort so important in the acceptance of the germ theory of disease?

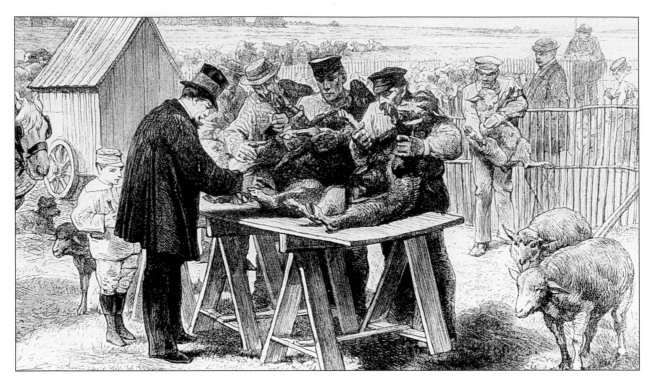

Figure 2.10

Louis Pasteur performing his anthrax vaccination experiment at Pouilly-le-Fort in 1881.

Pasteur and rabies

Soon Pasteur was to face the challenge of showing that his methods were a safe and successful way to protect people as well as animals from disease. He and his assistants had developed a technique of vaccination to protect dogs from rabies. Eventually they had discovered that they could prepare an effective vaccine by drying out parts of the spinal nerves from rabbits which had been infected with rabies.

When rabid dogs bite people they develop the fatal symptoms of hydrophobia which lead to an agonising death. In 1885 a young boy in Alsace was badly bitten by a dog with rabies. His mother had heard of Pasteur and his work and she travelled with her boy to Paris as quickly as was possible to ask Pasteur to help her son.

Figure 2.11

Pasteur with English children who were sent to him for treatment after they had been bitten by dogs.

Pasteur and his assistants were very nervous of trying their treatment on a human being but they knew that if they did nothing the boy would die. Fortunately, he survived. News of this success brought many others to Paris in search of a cure (Figure 2.11). Not all those who were treated survived, often because they were vaccinated too long after they were bitten. In the early months only 10 people died out of the 1726 people treated at a time when 16 out of every 100 people died when bitten by dogs with rabies.

9 What are the similarities and differences between Jenner's vaccine for smallpox and Pasteur's vaccines for anthrax and rabies?

10 Pasteur's opponents argued that Pasteur was risking giving people rabies by his injections and this was not justified because most people bitten by rabid dogs did not die. What is your view of the argument that Pasteur should have spent more time developing the treatment before trying it on people?

11 At the time that Pasteur was studying rabies there was no other research centre in Europe that could repeat his work. Why did this make it harder for Pasteur to convince other scientists and fellow doctors that his rabies vaccination was effective and safe?

Immunisation and TB

Every bacterium is different and it has taken years of research since Pasteur's time to develop safe and effective vaccines for a wide range of diseases.

No vaccine yet exists which is fully effective against TB. The BCG (*Bacillus Camille Guerin*) vaccine was invented in 1921. This is a weakened (attenuated) non-human strain of TB. It is very useful in preventing some forms of TB. In most parts of the world babies are vaccinated soon after birth. At the moment this is a successful programme with about 85% of babies receiving the vaccine worldwide. Immunisation helped in the final lowering of the number of cases of TB in countries like the UK as well (Figure 2.5). In the UK in recent years the BCG vaccine has been given to school children between 10 and 13 years old, and to adults such as teachers and health-care workers who meet a lot of people in their work.

Immunisation and influenza

The disease

Influenza – more commonly referred to as 'flu – is a relatively common respiratory disease caused by the influenza virus. There are several strains of the virus. The disease is highly infectious and has a very short incubation period. The symptoms of influenza include fever, often accompanied by shivering and sweating, feeling very unwell and unable to do anything, loss of appetite, aching muscles and painful joints. Simple influenza lasts for about 5–7 days before the fever goes down and convalescence begins but the exhaustion which follows can last from 6–12 weeks, even in patients without secondary bacterial infections.

Influenza infects the cells lining the tubes leading to the lungs, causing them to die. This leaves the airways open to infection, and many of the deaths associated with influenza are from severe secondary bacterial infections on top of the original viral invasion. This makes influenza more hazardous than other respiratory diseases, such as the common cold. The people most likely to die as a result of influenza are the elderly and anyone who is prone to asthma or heart disease.

Influenza spreads very quickly and there are often major outbreaks affecting thousands of people – these are known as epidemics. Three massive outbreaks (pandemics) which have affected much of the world have been recorded in this century alone, starting in 1918, 1957 and 1968. During the 'Spanish flu' pandemic in 1918–1920 at least 20 million people died from influenza.

The treatment for influenza is rest, warmth, plenty of fluids to avoid dehydrating and mild painkillers – although there are drugs which will ease the symptoms there is no drug yet which will cure the disease. If secondary bacterial infections set in, then antibiotics can be used to combat them.

Figure 2.12

Soliders returning home after the First World War. The 1918 outbreak of flu hit populations (soldiers and civilians alike) exhausted by the rigours of the First World War. It caused more deaths in a few months than had occurred throughout the whole of the war.

Questions

12 Why is it possible to catch influenza more than once despite the body's immune system?

13 Influenza vaccines are widely given only to certain groups of the population, particularly the elderly, rather than to everyone. Why do you think this is? Is it acceptable to limit access to a vaccine which could save lives? On what basis should such a decision be made?

Can influenza be controlled?

Each year the various strains of the virus are subtly different as the proteins on the surface of the virus change. The change is usually quite small, so having influenza one year leaves people with some immunity against infection for the next. But every so often there is a major change in the surface proteins - and this heralds a major outbreak as no one has any 'almost right' defences ready.

There are influenza vaccines, but because of the changing nature of the virus, the vaccine has to be different each year. A cocktail is made up of the strains of the virus thought most likely to cause disease in any one year, and this vaccine is then made available to those in high-risk groups such as the elderly and medical workers. The great fear is that a new and very different strain will appear again, and without an effective vaccine, millions of people worldwide could die.

The WHO is responsible for a worldwide surveillance network with over 100 centres for monitoring influenza outbreaks. This network keeps track of the strains of the influenza virus circulating in the world and recommends the appropriate composition for the vaccine each year.

Vaccination policy and safety

Vaccination gradually became more popular after the demonstration by Jenner that the procedure could be safe and effective. Smallpox was such a fearful disease that people were willing to overcome their fears (see Figure 2.8). By 1853 the smallpox vaccination had become widely available in the UK and was made compulsory for infants before they were 3 months old. The proportion of children vaccinated, however, was never close to 100%. Compulsory vaccination ended in 1946.

Smallpox vaccination sometimes caused serious side-effects, but so long as there was a significant chance of catching the disease most people felt that the risk was worth taking. By 1971 the chance of catching the disease in Britain was so low that routine vaccination for children was no longer recommended.

Immunisation is not only of value to individuals. It can help to prevent the spread of disease so long as a high enough proportion of people are immune. A small number of vulnerable people are effectively protected if they only meet people who cannot catch the disease and pass it on. The benefits of mass vaccination programmes are, however, put at risk if too many parents are unwilling to have their children vaccinated.

There is no such thing as a perfect vaccine which protects everyone who receives it and is entirely safe. There are three questions which a parent might ask before allowing a child to be vaccinated:

- Is the vaccine really effective in preventing disease?

- What are the possible side-effects and what is the chance that my child will be affected?

- What will the authorities do to compensate if my child suffers lasting damage from a vaccination recommended as part of public policy?

Some effective vaccines can produce side-effects which are not serious and which clear up quickly. In some cases the side-effects are not caused by the vaccine but are the result of human error or are just coincidences. Where there is a possibility of more serious side effects it is seldom possible to predict for sure which children are likely to be affected.

MMR vaccine and autism

The Department of Health recommends that infants receive MMR vaccination against mumps, measles and rubella. These are serious diseases which can cause life-long damage and even death. The general view is that the safest way of protecting children is with a single dose of a combined vaccine instead of three separate injections. Over 250 million doses of the MMR vaccine have been injected into infants in Western Europe since 1988 without evidence of dramatic side-effects. In the UK about 600 000 children receive the injection every year (Figure 2.13).

In 1998 Dr Wakefield and others at the Royal Free Hospital published a paper reporting that they had found an association between MMR vaccination and the onset of autism. The team studied 12 children who

Discussion points

Parents are encouraged to weigh up the advantages and disadvantages of vaccination taking into account the benefits and risks both for the individual child and for the community as a whole. Is it reasonable to expect parents to make the decisions about vaccination? Where can they turn for trustworthy information and advice?

Do the benefits of mass vaccination for the community as a whole ever justify compulsory vaccination for all children?

Figure 2.13

Bottles of MMR vaccine.

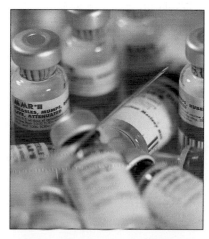

14 What do the two studies carried out at the Royal Free Hospital show about the importance of publishing scientific results?

15 The first study was well reported and worried a lot of parents. Was it right to publish new findings of this kind before they had been thoroughly checked and confirmed by other studies?

16 Were the scientists who published the first report right to mention the possibility of an association between MMR vaccination and autism? Are the two pieces of research conflicting?

17 After the first study reported in 1998, why did doctors continue to recommend MMR vaccination?

18 How did the second team of scientists account for the '5-month peak' and how did they check that it was due to coincidence rather than cause and effect?

19 Did the findings of the second study prove conclusively that the hypothesis connecting MMR and autism was incorrect?

20 The authors of the second report were asked: "Surely if there's even a one in a million chance that MMR vaccination could lead to autism then the vaccine is not safe?". What answer would you give to this question?

Discussion point

What was it about Pasteur's character and methods (as described in this chapter and chapter 1) which meant that he played such an important part is winning acceptance for the germ theory of disease?

had started to develop normally but then had suddenly developed the symptoms of autism. The parents said that in eight cases the problems had arisen soon after MMR vaccination. The doctors went on to examine another 40 autistic patients and found similar symptoms. In their report they pointed out that their work did not prove a link between the MMR vaccine and autism but they called for more studies.

These findings were widely reported and they worried parents. Doctors were concerned that fewer parents would have their children vaccinated putting their children at a much greater risk of catching serious infectious diseases. Doctors warned against placing too much confidence in one small-scale study. They pointed out that it is very risky to interpret case reports as suggesting a causal link when the association between the alleged cause and the symptoms may just be a coincidence.

Another team of researchers at the same hospital set out to test the hypothesis that MMR vaccination might cause autism by studying the case notes of 500 children diagnosed with autism who were born between 1979 and 1992. In their paper, which was published in 1999, they came to the conclusion that there was no connection. They noted that the trend in the incidence of autism did not change when the MMR vaccine was introduced in 1988. Furthermore, after 1988 the uptake of MMR vaccine had been constant while the incidence of autism continued to rise at the same rate as before, probably due to better diagnosis. Also they found no connection between the time when parents and doctors first became concerned about a child's development to see if this bore any relationship to when the child was vaccinated.

The doctors did note a '5-month peak' because quite a number of parents reported that they were first concerned about their child's health when the child was 18 months old, regardless of whether or not their child was vaccinated. This was 5 months after the age when most children receive their MMR vaccine. However the data showed that the onset of autism was not delayed if children were vaccinated later.

What makes science successful?

Science has produced the most reliable knowledge we have of the natural world – knowledge which is a basis for action such as discovering methods for preventing and curing disease. This has led people to try to explain what it is about the method of science that makes it so successful. Three of the more important accounts are summarised in Figure 2.14.

Figure 2.14 Three explanations of how science 'works' to produce reliable knowledge.

Explanation 1	Explanation 2	Explanation 3
Knowledge is reliable because it is grounded on data. If scientists have data on enough cases, of a wide enough variety of types, then they can generalise reliably from this. The method is called inductive reasoning. However, it has proved impossible to describe reasoning strategies that would guarantee reliable generalisations. Although much knowledge does seem to be based on this sort of reasoning, it cannot be shown to always work – even if a set of rules is followed.	Science is the method of bold conjectures, followed by strenuous attempts to refute (or falsify) these. The important thing about the method of science is not how generalisations and theories arise but how we test them. A scientific theory should make precise predictions, so that they are open to falsification. Only in this way can our knowledge grow. This view is associated with the name of the philosopher Karl Popper. However, in practice, scientists seem happier if a test is in agreement with their theories. And if an unexpected result does arise, it has often proved better simply to note it and hope that scientists will later find ways to account for it.	Most of the time scientists work within the current framework of ideas (or paradigm), which they have been taught. It is only because scientists are trained to work within such tightly defined frameworks that anomalies (unexplained results) are recognised. Usually these are resolved within the existing framework. So most 'normal science' is a kind of 'puzzle solving'. Occasionally anomalies build up which suggest that the current framework is wrong. After a period of uncertainty, a revolution in thinking occurs, and people quite quickly switch to the new framework. The decision to switch is based on evidence, but is also, to some extent, a decision of the scientific community in which tacit factors play a part. This view is associated with the historian Thomas Kuhn.

Although these are theories about science, they are not scientific theories. And so there is no consensus about which is correct. There are objections to all of them – as well as points that many people agree with. They provide useful perspectives, however, for thinking about some of the case-studies of scientific change and discovery described in this book.

Review Questions

21 Summarise the methods of preventing disease described in this chapter. Which methods depend on some kind of medical treatment? Which methods require non-medical interventions?

22 Refer to Figure 2.14. Illustrate aspects of the three explanations for how science 'works' with examples taken from Chapters 1 and 2. Give one or more examples of:

a a generalisation based on a collection of data,

b an explanation arising from conjecture and creative imagination,

c a prediction based on theory and tested by experiment,

d an unexpected observation which did not fit with the accepted theory but which was ignored,

e puzzle solving within an established framework of normal science,

f a revolution in the accepted framework of scientific thinking.

Chapter 3

Medicines to treat disease

The issues

Most people welcome the development of new drugs, and better treatments for the sick. Healthcare and the development of new treatments is very expensive and the benefits have to be weighed against the costs. The demand is always greater than the health service can provide. This means that doctors have to make judgements about the treatments they prescribe, taking into account, among other things, the chances of successful cures, the extent to which the treatments will improve the quality of life of patients and their cost.

A growing worry is that the tried and tested antibiotics are beginning to fail as more and more bacteria develop resistance. Overuse of antibiotics in medicine and agriculture has accelerated the evolution of resistant bacteria which are no longer killed by the drugs.

What this tells us about science and society

Innovations based on science (such as new drugs) can help to improve the quality of life of many people but at the same time there may be unintended and undesirable side-effects. Controlled clinical trials are a crucial method of obtaining valid evidence to support the fact that a new approach is an improvement. By using a control group, scientists can gather evidence to prove that one specific factor makes a difference.

Testing new approaches before they are put into practice raises ethical issues: about the use of animals in testing, and about including or excluding people from trials whose outcome may be beneficial or have harmful side-effects. Yet without rigorous testing, we cannot make improvements which may save lives in the future.

Scientific studies can explore alternatives to using animals for tests. They can also help to weigh up the likely medical benefits and the risks involved in not doing any tests. But science alone cannot resolve the ethical issues which call for judgements based on principles such as respect for living things and fairness.

Figure 3.1

Pharmacist preparing a patient's prescription in a hospital pharmacy.

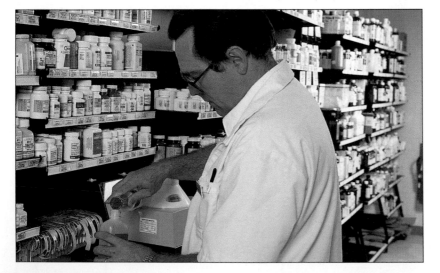

Drugs and medicines

Plants are the world's chief source of drugs. They provide most of the effective drugs in traditional medicines as used by 80% of people in the world. Western pharmacology recognises 7000 plant-based medical compounds now used in about one-in-four drugs in western medicine. Yet of the 250 000 flowering plants in the world only 5000 have been tested for useful drugs in laboratories.

The rosy periwinkle, for example, is a small plant with a long folk history (Figure 3.2). In the 1950s its reputation as a treatment for diabetes led Canadian scientists to investigate the plant for substances which could be used in place of insulin. Instead they discovered that extracts of the plant had powerful anti-cancer properties. Mice suffering from leukaemia and treated with rosy periwinkle extract lived longer then untreated leukaemic mice. Research continued and in 1958, chemists successfully isolated the substance active against the leukaemia cells. The substance, called vinblastine, was first tested in 1960 on a patient with a large cancerous tumour. The tumour rapidly disappeared. For more than 2 years the patient was free of cancer but then the symptoms reappeared. Fortunately, by then other substances had been isolated from rosy periwinkle and one, vincristine, brought the disease under control once more. Although the cancer cells had developed resistance to vinblastine, the resistance was not carried over to vincristine.

The pharmaceutical industry

The development of the modern pharmaceutical industry has been based on remarkable developments in techniques of analysis and synthesis which now allows chemists and biochemists to model and manipulate the detailed structure of complex molecules.

Until the mid-1930s the pharmaceutical industry was small, producing mainly simple chemicals and botanical extracts. Local pharmacists mixed the ingredients to make liquid mixtures, pills and ointments. Most of these drugs were not cures. Doctors prescribed them to relieve the symptoms while relying on the immune systems of their patients for the cure. Today we continue to take mild painkillers, cough mixtures and apply ointments in a similar way.

One or two of the older medicines were effective. Doctors could prescribe vitamins for patients suffering from some deficiency diseases. Digitalis from foxgloves was available to treat heart failure. Quinine was an anti-malarial drug which had been first isolated over a hundred years earlier from the cinchona tree in the tropical rainforests of the Amazon where its effects had been known by the peoples of the forest for centuries. More recently in 1922, the Canadian doctor Frederick Banting and a young medical student, George Best, had successfully isolated and tested insulin from the healthy pancreases of dogs and then cows. This drug was being used worldwide to treat diabetes within a few months of their first successful tests.

Figure 3.2

The Madagascar periwinkle (Catharanthus roseus) is a source of anti-cancer drugs such as vinblastine and vincristine. The discovery of these drugs illustrates the importance of listening to and understanding the ways the rainforest peoples practise their traditional medicine.

Key terms

Drugs are the active ingredients in medicines used for the treatment, relief or prevention of disease. People also take drugs for pleasure, stimulation and relaxation.

Medicines normally consist of one or more drugs mixed with some inert materials combined in a way which makes the treatment available as pills to swallow, ointments to rub onto the skin, powders or vapours to inhale, solutions to inject and drops for the eyes or ears (Figure 3.3).

The **pharmaceutical industry** is the part of the chemical industry which makes drugs and medicines.

Chemotherapy is the use of chemicals to treat disease.

Figure 3.3

An assortment of medicines which are past their sell-by date and have been collected for disposal. Note the range of types of formulation (such as pills, capsules, ointments, drops, injections, inhalers, powders).

Figure 3.4

Paul Ehrlich and his colleague Sahachiro Hata.

Figure 3.5

Red and white blood cells viewed through a light microscope. The white cells have taken up the colour of a dye used to stain the cells.

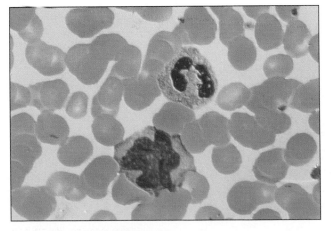

However, the revolution in the use of chemicals to treat disease was marked in 1935 by the launch of sulphonamide drugs. This was followed in the 1940s by the discovery of antibiotics. These discoveries transformed the pharmaceutical industry. Companies began to build up research teams engaged in systematic studies to develop new drugs. The companies have grown as the costs of research have risen so that most of them are now large, international businesses.

The beginnings of chemotherapy

Paul Ehrlich (1854–1915) is often regarded as the father of modern chemotherapy (Figure 3.4). Ehrlich's idea was that it might be possible to make chemicals to kill the microbes which cause disease without harming other living cells.

Disinfectants destroy microbes outside the body but they cannot be used inside the body because they are so toxic. The problem is to find chemicals which will destroy an internal infection without harming the patient too much. The trick is to find a chemical 'magic bullet' which will destroy the cause of disease but leave healthy tissue unharmed.

It seems that Ehrlich's thinking was influenced by his interest in dyestuffs. He preferred to experiment with dyes in the chemistry laboratory when he should have been studying medicine at the University of Breslau. This meant that he took a long time to pass his medical exams.

The first synthetic dye was made by William Perkin in England in 1856. German chemists visited London and Manchester to learn about dye manufacture. Soon the industry began to develop rapidly in Germany. Methods were discovered of making new dyes based on chemicals from coal tar.

Among the new dyes were magenta and methylene blue. In the 1870s, Robert Koch developed the methods used to study bacteria. He used magenta and methylene blue to stain bacteria on glass slides so that they could be seen under the microscope. Ref to pages 7–8 Paul Ehrlich was one of Koch's best assistants. He was particularly interested in the idea that dyes could be used selectively. Some dyes take well on wool but not on cotton. Certain dyes will stain some animal cells but not others.

Ehrlich showed that selective dyes could be used to classify blood cells (Figure 3.5). He also showed that if methylene blue is injected into an animal it will dye nerve cells but not other parts of the body. With this idea in mind, Ehrlich began his hunt for 'magic bullets'. He thought it might be possible to inject dyes into a patient which would kill microbes but leave healthy parts unharmed.

During a long period of research he investigated the effect of new synthetic dyes on blood parasites called trypanosomes. Sleeping sickness is one of the diseases caused by these parasites. Ehrlich showed that the dyes were effective in killing the parasites in infected mice. Unfortunately he had not found a 'magic bullet' because they also poisoned the animals.

After his lack of success with coal-tar dyes Ehrlich decided to study arsenic compounds. The dyes he had been using were nitrogen compounds. Arsenic is in the same group of the Periodic Table as nitrogen so Ehrlich thought that arsenic compounds might be worth investigating.

Over 600 arsenic compounds were made and tested with no positive results. Ehrlich decided to try every one again. In 1909 he was working with a Japanese colleague, Sahachiro Hata. Together they found that the six hundred and sixth compound hit the target. Its effectiveness had been missed by a technician during the first series of trials.

The 'magic bullet' was found to be effective against trypanosomes in mice. Unfortunately, it had no effect on the parasites which cause sleeping sickness in human beings.

Ehrlich now decided to try it on other microbes. He used it with the bacterium which causes syphilis. Ehrlich found his arsenic compound cured syphilis in rabbits. He later found it cured the disease in humans too. He had discovered the first synthetic chemical to control a parasitic disease. He called the new drug 'Ehrlich 606' after the long struggle for success. It was patented in Germany and sold as 'Salvarsan'.

Prontosil

Another drug was discovered as a result of research into dyes in the 1930s. A new red coal-tar dye had been made and a sample was passed to Gerhard Domagk. Domagk was director of a laboratory investigating the value of dyes as drugs.

Domagk was interested in the new dye because it stuck strongly to wool. Wool is a protein, and this suggested to him that the dye might stick strongly to the proteins of bacteria. When tested on mice the dye was found to be very effective against a variety of bacterial diseases.

The first person to be treated with the new drug was Domagk's daughter, Hildegarde. She picked up a serious infection by accident in his laboratory. Her life was in danger. As a last resort Domagk suggested treatment with the red dye. It was successful and her life was saved.

The red dye was the first of the sulphonamide drugs. It was called 'Prontosil'. It became famous when it was used to fight an outbreak of childbed fever at Queen Charlotte's Hospital in London.

Three years later two chemists working for a British drug company discovered another successful sulphonamide drug after a long series of trials with nearly 700 other compounds. After tests on mice the drug was tested on a farm worker in Norfolk who was dying of pneumonia. He recovered in a few days. They later discovered that the drug worked by stopping bacteria from growing and multiplying. It was an anti-bacterial but it lacked the full power of the new generation of antibiotics developed as a result of Alexander Fleming's unexpected observations.

Questions

1 Why is it so difficult to find a chemical which will kill the microbes which cause disease without harming the healthy parts of the body?

2 How did the discovery of synthetic dyes help Koch in his research to identify the bacteria which cause disease?

3 What was the 'model' which Ehrlich used in his thinking about chemicals to cure disease? How did the context in which he was working help him to develop this model?

Questions

4 Why did Domagk decide that a dye which was stuck strongly to wool might also be attracted to the proteins of bacteria?

5 The 'magic bullets' such as Prontosil became less important in the treatment of bacterial disease after the 1940s. Suggest a reason for this.

6 All chemotherapy is likely to have some unpleasant side-effects. Suggest a reason why?

The discovery of antibiotics

A slice of bread left open to the atmosphere for a few days becomes a battleground for microbial warfare. In Figure 3.6, notice the splashes of colour spread over the bread's surface. Each colour is a colony of a particular species of bacterium or mould. Once a colony has established itself it prevents colonies of other species from trespassing on its territory by releasing a substance which kills the invaders. In other words the substance acts as an antibiotic.

One summer's day in 1928, the British bacteriologist Alexander Fleming was examining some Petri dishes in which were growing colonies of *Staphylococci* - the kind of bacteria which causes boils and sore throats. The colonies were several days old, and Fleming noticed that one of the dishes was contaminated with a mould. He was surprised to see that the colonies of *Staphylococci* growing near to the mould were dying. It looked as if the mould were producing a substance which killed the bacteria.

Figure 3.6

Different species of mould growing over the surface of a piece of bread. The fruiting bodies of Penicillium are blue while those of Aspergillus are green or yellow. The pin moulds such as Mucor form black fruiting bodies.

Key term

An **antibiotic** is an antibacterial substance produced by a fungus or other organism.

Fleming cultured the mould. He found that substances released into the culture solution could kill bacteria that cause human diseases, and that the solution was effective even when diluted. Fleming also injected the solution into mice and showed that it did not harm them. The mould was later identified as *Penicillium notatum*, and Fleming named the mysterious anti-bacterial substance it produced penicillin. By chance, he had found the first antibiotic but did not fully understand the importance of his discovery. At the time other scientists were indifferent to his work. They were concentrating on other ways of treating human disease.

There matters rested until 1938 when Howard Florey and Ernst Chain at Oxford University took up the investigation (Figure 3.8). They quickly established penicillin's effectiveness against different bacteria and that it was harmless when injected into mice. By 1941, the scientists had developed methods to produce enough penicillin for clinical trials. The mould was grown in culture using all kinds of vessels (including milk bottles) to contain the mould/culture mixture. About 100 litres of the

Figure 3.7

Fleming at work in his laboratory at St Mary's Hospital, London in 1909. It was nineteen years later that he discovered that secretions from the fungus Penicillium notatum *destroy colonies of the bacterium* Staphylococcus *species*.

mixture were needed to produce enough penicillin for one day's treatment of one patient. The supply was so short that penicillin was saved by extracting the drug from the urine of treated patients.

The trials proved penicillin's efficacy but by then the Second World War (1939–45) had begun. There was an urgent need to treat wounded soldiers. Untreated wounds often became infected with bacteria causing fatal diseases. Production of penicillin had to increase, and the work moved to the USA where large-scale manufacturing processes were developed. By 1944, the production of penicillin was sufficient to treat all of the British and American casualties that followed from the invasion of the European mainland to defeat Hitler's Germany. Today, the massive demand for penicillin worldwide is met by the development of new high-yielding strains of *Penicillium* mould.

Improvements in yields have also come through advances in genetic engineering. Penicillins are produced in huge containers (fermenters) which hold up to 200 000 litres of mould/culture solution. At the end of a production run, the mould is filtered off and the penicillin extracted from the solution.   *Ref to Chapter 7*

Biochemists have now discovered how penicillin destroys bacteria. The drug fatally weakens the cell walls of the bacterial cells so that they burst and die. Human cells do not have cell walls so they are not affected by the antibiotic. Penicillin is close to being the perfect 'magic bullet' which Ehrlich dreamed about. What he did not anticipate was the problem of drug resistance.

Figure 3.8

(a) *Ernst Chain checking the formulas for penicillin.* **(b)** *Howard Florey injecting penicillin into the tail of a mouse in the 1940s.*

Antibiotic resistance

With time, antibiotics become less effective because bacteria develop resistance. Following the introduction of penicillin in the 1940s, strains of bacteria have emerged which produce the enzyme, penicillinase, which breaks down the penicillin, making the drug ineffective.

In the case of penicillin, populations of bacteria always contain a few individuals with genes for penicillinase making them resistant to the antibiotic. These individuals survive the onslaught of penicillin when a patient is treated with the drug, and reproduce new individuals which inherit the genes for penicillinase. The offspring are resistant and, because bacteria multiply very quickly (in some species a new generation is produced every 20 minutes), resistance spreads quickly. To be effective, the dosage of drug has to be increased step by step until the drug becomes so ineffective or poisonous to the patient that an alternative has to be found.

Developing new drugs

The first step in the development of a new drug (Figure 3.9) is to identify a disease for which the existing treatments are inadequate or not available. Detailed research into the disease helps to pinpoint the processes in the body which could be affected by a drug.

The process of research and development is long and slow, so it costs a company a vast amount of money to bring a new drug to market. The company therefore takes out patents in many countries to protect its discovery and prevent other firms making and selling the drug except under licence. Patents only last a limited number of years so the company hopes to be able to sell enough of the drug to recover its costs and make a profit before the patent runs out.

The action of drugs is closely linked with the structure of its molecules. Nowadays a research team looking for alternatives for an existing treatment might start with the structure of an established drug and explore ways of changing it to produce a superior product. Much of the

Figure 3.9

Stages in the development of a new drug.

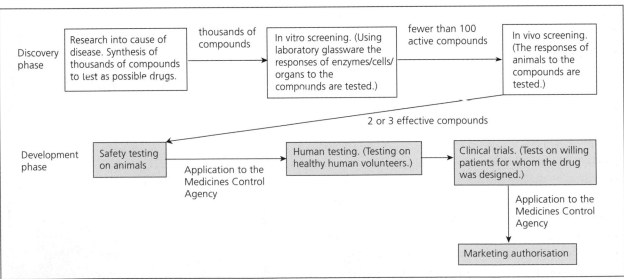

early planning can be done by computers modelling drug structures on screen and matching them, on a molecular scale, with likely sites in the body where they might be effective (Figure 3.10).

For a research compound to become a marketed product it has to cross three critical barriers: it has to meet a definite need for treatment; it has to be technically possible to develop and manufacture the product; and there has to be a big enough market to ensure that the development work will be commercially viable.

Figure 3.10

Computer design also speeds up the discovery of new drugs. Even very small changes in the shape of a potential drug molecule can improve its performance, and computers help scientists to screen huge numbers of variations, a few of which might make a big difference in the effectiveness of treatment.

Figure 3.11

A researcher taking a blood sample from the ear of a white rabbit to see if it is infected with the HIV virus which causes AIDS. This is part of a research project attempting to prevent the HIV virus from reproducing using anti-viral drugs and gene therapy.

Testing new drugs

All claims made for the medical benefits of a new drug and for its proposed use in treatments have to be supported by scientific evidence. Conventional medicines must normally satisfy the requirements for efficacy, quality and safety. Manufacturers are therefore obliged to carry out a rigorous testing programme before bringing a new drug to the market.

Drug testing begins in the laboratory on preparations of cells, tissues and even whole organs. This is called *in-vitro* testing. The preparations are placed in glass containers, each containing a solution of all of the substances needed to keep the preparation alive. The compound under test is added and its effects on the preparation observed.

When a compound has the desired effect *in-vitro*, it begins to look interesting. However, will it do the same in the patient? *In-vivo* testing in a living body is the next stage. The decision to proceed further with a promising compound is a crucial one, because the investment in time and money now becomes substantial. *In-vivo* testing begins in animals. Several years are spent looking at every aspect of the candidate drug (Figure 3.11).

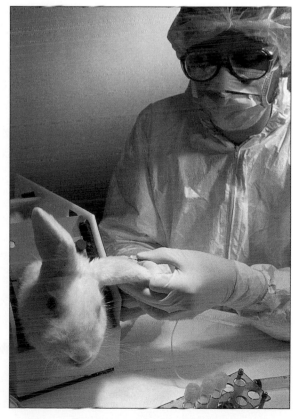

Figure 3.12

Laboratory mice are used for safety tests on drugs. The size and genetic make up of mice makes them suitable for medical and toxicity tests. Licences are issued to the individuals who are responsible for animal experiments.

Manufacturers have legal obligations to the animals used during trials of new drugs. Indeed Britain made history in 1876 when it became the first country to pass a law which controlled research with laboratory animals. Subsequently the law has been updated, so that in 1986 the *Animals (Scientific Procedures) Act* established three main areas of legal control over animal research:

- The people who carry out the research must be competent to do so (Figure 3.12). Competence to carry out *particular* types of experiment on *named* species of animal in the course of an *approved* project is established by the issue of a **personal licence**.

- The facilities for looking after animals before, during and after experiments should be suitable (Figure 3.13). A **certificate of designation** is only granted to scientific institutions where the places for keeping animals meet the necessary high standards.

- The likely benefits of the research should outweigh any possible distress to the animals. Only when the balance in favour of human welfare is established will a **project licence** be issued for the work to go ahead. Even then the law aims to protect animals from unnecessary distress. Scientists are required to prevent or relieve pain wherever possible.

About 85% of all experiments involve rats or mice. However, if dogs, cats, monkeys or apes are involved, the holders of the project licence must justify use of the larger animals.

The ethics of drug testing with animals

Most people agree that it is wrong to make animals suffer. Many also agree, however, that there are some circumstances in which the interests of animals may be outweighed by the interests of human beings. One example would be the use of animals to test new medicines, which many people find acceptable so long as everything possible is done to minimise pain and distress.

Figure 3.13

Transgenic sheep in a feeding pen. These lambs are offspring of ewes that have a human gene in their DNA. This means that they produce a protein in their milk that can be used to treat a hereditary deficiency disease. Wherever possible experimental animals are housed in groups to satisfy their behavioural needs as well as their physical needs.

The view of organisations which defend the use of animals in medical research:

Without this research, few of the drugs we now take for granted would now exist, and the chances of finding new or improved treatments in the future would be extremely limited.'

The view of an animal welfare organisation:

As long as animals continue to be used in experiments they must be given the maximum protection from pain and suffering, whatever the purposes for which they are used.'

The view of anti-vivisectionist organisations:

Animal experiments are both unjust and unnecessary and cannot be justified, for any reason.'

Figure 3.14

Contrasting views on the use of animals in medical research.

Others, however, hold the view that all use of animals for human beings for medical purposes is wrong, no matter how great the benefits. In the UK only a minority of people would like to see an absolute ban on drug testing with animals but some members of this minority hold to their views passionately and are prepared to fight to win the argument (Figure 3.15).

Question

8 How do your respond to the statements in Figure 3.14 and which is closest to your own view?

Figure 3.15

A women taking part in an animal rights demonstration in Coventry, 1995.

What is ethics?

Ethics is the branch of philosophy concerned with how we should decide what is morally wrong and what is morally right. An example of an ethical question is: 'Should we use non-human animals in medical experiments?'.

Answering questions of this kind is often far from easy. Science plays a part in answering ethical questions but is not enough on its own. So how do people carry out ethical thinking, or in other words, 'do ethics'?

The way ethics is done

Ethics is a branch of knowledge as are other subjects, such as science, mathematics and history. Ethical thinking is not wholly distinct from thinking in other subjects but it cannot simply be reduced to them. In particular, ethical conclusions cannot be unambiguously proved in the way that mathematical theorems can. However, this does not mean that all ethical conclusions are equally valid (that is, correct) any more than all scientific or

historical conclusions are equally valid. As a fairly safe rule of thumb, the first thing to do when deciding if something in science or technology would be ethically acceptable is to look at the likely consequences.

For example, to decide whether or not non-human animals, such as mice or chimpanzees, should be used in medical experiments, think first of all the consequences

- of using such animals; and

- of not using such animals.

In this particular case, obvious major consequences include: the likely medical benefits to humans from using animals in medical experiments and the likely harmful effects on the animals from their being used.

However, notice immediately how difficult it is to work out these consequences in any detail. Even if a careful review of the historical evidence shows how useful in the past medical experiments on animals have been for human health, this cannot accurately predict how useful medical experiments on animals will be in the future. It is because of this uncertainty that those who oppose the use of animals in medical experiments can maintain that animal experiments are no longer needed, and can be replaced by tissue culture, computer modelling and the use of human volunteers, while those in favour of the continued use of animals in medical research can dismiss these alternatives as insufficient.

Utilitarianism

No one believes that we can ignore the consequences of an action before deciding whether or not it is right. The deeper question is not whether people need to take consequences into account when making ethical decisions but whether that is all that they need to do. Are there certain actions that are morally required – such as telling the truth – whatever their consequences? Are there other actions – such as experimenting on people without their consent – that are wrong whatever their consequences?

Those who believe that consequences alone are enough to let us decide the rightness or wrongness of a course of action are called consequentialists. The most widespread form of consequentialism is known as utilitarianism. Utilitarianism itself exists in various forms, but it begins with the assumption that most actions lead to pleasure and/or displeasure. In a situation in which there are alternative courses of action, the desirable (that is the right) action is the one which leads to the greatest overall increase in pleasure.

Consider the question as to whether or not doctors should tell the truth to their patients. A utilitarian would hesitate to provide an unqualified 'yes' as a universal answer. Utilitarians have no moral absolutes beyond the maximisation of pleasure principle. Instead, it would be necessary for a utilitarian to look in some detail at particular cases and see in each of them whether telling the truth would indeed lead to the greatest net increase in pleasure.

Questions

9

a A utilitarian might argue that there is no logical reason to distinguish the pain and suffering felt by animals from that felt by human beings and that it would be wrong to weight animal suffering less heavily than human suffering. What do you think?

b Can the benefits of testing new drugs for use by doctors or vets produce such benefits that they outweigh the distress caused to animals while the drugs are being developed?

10 Is it logical for someone who eats meat and wears leather shoes to object to the use of animals for testing drugs?

Ethical principles

The major alternative to utilitarianism is a form of ethical thinking in which certain actions are considered right and others wrong in themselves, that is intrinsically, regardless of the consequences. Consider again, for example, the question as to whether scientists should use non-human animals in medical experiments. Some people believe that this is unacceptable partly because the animals may suffer but more fundamentally because the animals used in such research have not given their consent. After all, scientists do not use children to see whether drugs are toxic, so why is it acceptable to use rats?

Someone arguing in this way might maintain that non-human animals have certain 'rights', just as humans do. On the other hand, there are those who argue that it is acceptable for us to use non-human animals for such tests precisely because humans have a right to use animals for their own purposes. After all, people do not allow animals to choose whether they are prepared to be kept on farms or as pets so why should we be required to use animals for research only if such animals have given their consent?

There are a number of possible intrinsic ethical principles most of which are concerned with rights and duties of various kinds. Perhaps the most important intrinsic ethical principles are those of autonomy and justice. People act autonomously if they are able to make their own informed decisions and then put them into effect. At a common-sense level, the principle of autonomy is the reason why people should be provided with access to relevant information, for example, before consenting to a medical procedure or taking part in a medical trial.

Autonomy is concerned with the respect due to individuals. Justice is about fair treatment and the fair distribution of resources or opportunities. Considerable disagreement exists, though, as to what precisely counts as fair treatment and a fair distribution of resources. Answering ethical questions often isn't easy.

Clinical trials

The design of trials

Only when the results of the animal studies have been approved by the *Medicines Control Agency*, can clinical trials of the drug on humans begin (Figure 3.9). Phase I trials are short-term studies of the effects of the drug in a small number of healthy volunteers. In phase II, the programme with healthy volunteers is extended and short-term studies in several hundred patients begin. Phase III involves large studies in several thousand patients. The clinical trial programme may take as long as 8 years to complete.

The best form of clinical trial is a 'double blind study' in which the drug to be tested is given to some patients but not to others. People are only included in the trial if they agree to take part.

The patients who are not given the drug are the control group against which the results of giving the drug to the other patients can be compared. People in the control group receive a placebo: an

> ## *Discussion points*
>
> Imagine and describe a situation in which a doctor might be torn between telling the truth to a patient and not telling it.
>
> Is it fair that people living in rich countries have far more resources than people in poorer countries? If you think the answer to this question is 'yes', explain why. If you think the answer is 'no', how should a fairer distribution of resources be brought about?
>
> Some argue that the use of animals for medical research violates their basic rights as living things. Animals, they would argue, do not exist to be exploited by human beings. In your view do animals have rights and if so, what are the implications for the testing of drugs?

inactive substance that does not effect them. Obviously the active and inactive medicines must look similar and the placebo is sometimes called a dummy.

People are chosen for the control group by allocating at random a sealed envelope to each patient who has been accepted on to the trial. Each envelope contains a number. A patient's envelope is opened at the control centre for the trial and the number determines whether the patient receives the drug or the placebo. Only the controller of the trial holds the code to the significance of the numbers. Neither the patients nor the doctors know who has been given which treatment. In this sense, both are *blind* – hence the name 'double-blind' studies. The method effectively prevents bias creeping into the results, which would otherwise happen if doctors chose which patients were to receive the drug. Figure 3.16 summarises the sequence of events.

Figure 3.16

Randomisation and 'blindness' are essential components of clinical trials.

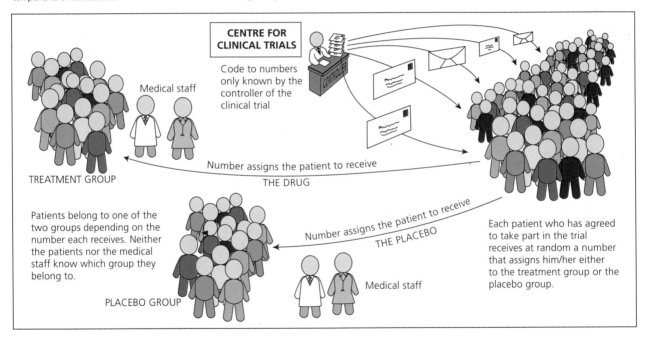

CENTRE FOR CLINICAL TRIALS

Code to numbers only known by the controller of the clinical trial

TREATMENT GROUP

Medical staff

Number assigns the patient to receive THE DRUG

Patients belong to one of the two groups depending on the number each receives. Neither the patients nor the medical staff know which group they belong to.

Number assigns the patient to receive THE PLACEBO

PLACEBO GROUP

Medical staff

Each patient who has agreed to take part in the trial receives at random a number that assigns him/her either to the treatment group or the placebo group.

In the UK the data from clinical trials is sent to the *Medicines Control Agency* who appoint an independent panel of experts to evaluate the drug. If the members of the panel conclude that the drug provides more benefit to patients than the placebo alone, then the manufacturers can apply for a product licence and begin to market the drug (Figure 3.17). In the USA permission to manufacture and market a new drug is given by the *Food and Drug Administration* (FDA). Other countries also have their own regulatory bodies.

Medical ethics committees

The task of an ethics committee is to decide if the potential benefit of a clinical trial is worth the risk to the volunteers. For this reason the committee should be independent of the company producing the drug and the research team carrying out the trial. The top priority is the health and safety of the volunteers.

Questions

11 What do you understand by the term 'placebo effect'? What might account for this effect?

12 Why is important that neither the patients nor the medical staff know which people are receiving the drug being tested? How might the results be affected if the distribution of the drug to be tested and the placebo were not kept secret?

An ethics committee for clinical research typically consists of two doctors (one from general practice), two people from professions allied to medicine (such as nursing or pharmacy) and an equal number of lay people (ideally including a lawyer and someone from a religious organisation). Ideally the committee should mirror the make up of the local community too in terms of the sex and ethnic background of its members.

Ethics committees advise on the recruitment of volunteers for phase I and phase II trials on healthy patients so that they are not sought out from groups which seem too vulnerable. The ideal volunteer is someone who would like to earn some money for taking part in the trial but is not desperate for it.

Ethics committees do not make recommendations on the size of the fee paid to volunteers but they may comment if they think that the fee is too high and likely to attract unsuitable people to the trial. The fee should be large enough to compensate volunteers for the inconvenience and discomfort involved in taking part in the trial, but should not be large enough to essentially bribe the volunteer to do something he or she would rather not do.

Members of an ethics committee want to see that volunteers are being thoroughly screened so that unsuitable volunteers are not included. A trial, for example, should not involve people who are taking or have recently taken other drugs which might interact with the drug to be tested.

Discussion points

Why include lay people on ethics committees? Why not leave the decisions to experts?

Is it right that healthy people are paid to take part in the early phases of clinical trials? Who would you expect to be willing to volunteer to take part in such trials?

Figure 3.17

This young volunteer is an asthmatic taking part in a clinical trial of a new drug. The vapour coming from the pipe encourages deep breathing. The computer screen displays measurements of the concentrations of oxygen (blue, top) and carbon dioxide (yellow, bottom) in the blood.

New drugs to treat TB

The discovery of penicillin stimulated a search for other new antibiotics. Soon scientists in the USA discovered streptomycin which promised to provide an effective treatment for TB. The British government bought enough of the drug to treat up to 200 patients for a cost of £150 000 which was an enormous sum in the 1940s.

Not many patients could be treated and it was important to find out just how effective the new drug could be. So Austin Bradford Hill, a distinguished statistician, was brought in to design a randomised clinical trial. One hundred and seven patients were monitored in hospitals in London, Wales, Scotland and Yorkshire. Streptomycin was given to 55 patients and the other 52 received the only other available treatment, bed rest.

Questions

13 What difficult decisions arise when selecting very sick people for a clinical trial of a new drug which may save the lives of those treated?

14 New drugs are always expensive but they are popular with doctors and patients because of their benefits. The cost may mean that money is not available for less glamorous, cheaper but worthwhile treatments. Who should decide whether or not the health service should spend large sums on new drugs especially at an early stage when their benefits may still be uncertain? What evidence is required for a rational decision?

A particular patient's treatment was decided at random by the equivalent of tossing a coin. At the end of 6 months, 14 patients given bed rest had died and four had shown a considerable improvement. Of those given streptomycin only four had died and 28 were very much better.

For a while streptomycin was widely used in the treatment of tuberculosis, but it has harmful side effects so it has been superceded by new antibiotics. The drug can affect the nerves from the ears to the brain leading to disturbed hearing, possible deafness and problems with balance.

TB on the rise

Each year more people are dying of TB. New outbreaks have occurred in Eastern Europe where TB deaths are increasing after 40 years of steady decline. The incidence of TB is also rising in south-east Asia and sub-Saharan Africa.

One factor speeding up the spread of TB is HIV infection. HIV weakens the immune system (Figure 3.18). This means that someone who is HIV positive is very much more likely to develop the symptoms of TB if they are infected. TB is now the leading cause of death for people who are HIV-positive and one-third of AIDS deaths worldwide are the direct result of tuberculosis.

HIV and AIDS

Figure 3.18

AIDS (acquired immunodeficiency syndrome) is a new human disease caused by HIV (human immunodeficiency virus).

(ACQUIRED) – something gained or caught

(IMMUNE DEFICIENCY) – destruction of the immune system so that the body cannot defend itself against infection

(SYNDROME) – the set of symptoms associated with a disease

AIDS was first recognised in the USA in 1981 among homosexual men who previously had enjoyed good health. The disease was already widespread in Africa and is now worldwide.

HIV attacks white cells (so-called T helper cells) in the blood which controls the immune system. Normally, the immune system keeps most of us healthy for most of the time. Its destruction by HIV exposes infected people to a variety of fatal diseases.

Infection with HIV does not necessarily lead to full-blown AIDS. People infected by the virus who do not display the symptoms of AIDS are described as being HIV positive. This is because tests on their blood detect antibodies to HIV.

Full-blown AIDS develops once HIV has done severe damage to the immune system. This allows other illnesses to develop including TB, pneumonia, skin cancer and brain damage.

Another serious problem is that badly managed TB treatments are threatening to make TB incurable. Until the discovery of antibiotics there were no drugs to cure TB. Now there are strains of the bacterium which are resistant to all the main anti-TB drugs. The development of drug resistance is accelerated if doctors prescribe the wrong treatment or patients do not take the full course of drugs and stop taking their medicine once they feel better.

Effective TB control

The World Health Organisation (WHO) has developed a treatment strategy for detecting and curing TB which is very successful if applied systematically. Millions of people have now benefited so that in much

Question

15 More and more people are travelling internationally on business and for holidays. There are many people who have been displaced from their homes by war, famine, natural disasters or the search for work. Why does movement of people help the spread of TB and make it more difficult to control the disease?

of China cure rates among new cases are 96 per cent while in Peru the programme has cured 91 per cent of cases (Figure 3.19).

The elements of the strategy are:

- political commitment,
- microscopy services to detect the TB bacteria in samples of saliva,
- drug supplies,
- a drug regime which has proved effective,
- monitoring of patients to observe and record patients

following the full course and taking the correct doses of drugs.

Figure 3.19

A nurse visiting a tuberculosis patient in Delhi, India, to make sure that she continues to take the treatment.

Drug resistance develops rapidly in tuberculosis if any one drug is used alone, so the initial phase of treatment today is designed to bring the disease under control as quickly as possible thus cutting the danger of resistance developing. Then a continuation phase of treatment starts once tests have discovered which drugs are most active against the strains of the TB bacterium affecting the particular patient. This continuation treatment may have to continue for many months.

The treatment of TB using combinations of drugs depends on the notion that bacteria resistant to one of the drugs are unlikely to be resistant to the others in the drugs cocktail. So far the majority of deaths from multi-drug resistant tuberculosis have been of patients who were also infected with HIV. However, the threat of multiple resistance to people with tuberculosis but without the complications of HIV infection is very real. Already there are reports of strains of bacteria resistant to all of the known anti-TB drugs.

Anti-viral drugs

At present there is no drug treatment available for viral diseases as powerful in its effects as antibiotics are against bacterial diseases. In other words, the anti-viral equivalent of penicillin has still to be discovered. However, new substances which are at the research stage or just becoming available may hold the key to an anti-viral future.

Very few drugs are effective against viral diseases such as influenza, although amantidine may be useful for patients who are at serious risk from other infections. The drug prevents the influenza virus from reproducing in the cells of the mucous membrane. Another drug called 'Zanamivir' is also available. It helps reduce the length of time the patient suffers from influenza symptoms by about 24 hours.

You might think that an anti-viral future is just around the corner. However, the make-up of most viruses is constantly changing (for

Discussion point

From the point of view of public health it can be argued that poorly supervised treatment of TB is worse than no treatment at all. Do you agree?

Questions

16 Explain the importance of political commitment in any countrywide programme to control TB.

17 Patients have to be watched while taking their drugs during the TB treatment programme recommended by WHO. Why is this necessary? Do people responsible for medical care have the right to insist that patients are kept under surveillance during treatment?

18 What aspect of human nature and what pressures on people have contributed to the development of resistance to the drugs efffective against TB? What might have been done, worldwide, to prevent resistance developing? What are the lessons for the future?

example the pattern of coat protein frequently alters). The changes are the result of mutations in the viral genetic code. Anti-viral drugs effective against a particular form of a virus can quickly become useless against the different mutated version. New drugs may be developed to cope with the different version, but the threat of future mutations keeps research scientists hard at work trying to stay one step ahead.

Ref to page 22

Recent research has developed new anti-influenza drugs (Relenza and GS4104) which appear to be very effective in stopping the virus from multiplying. These drugs target a feature of the virus which is unaffected by mutations and which is essential if each new generation of viruses is to attack and infect more cells.

Review Questions

19 There are two ways of fighting disease: preventive medicine to stop people getting ill and therapeutic medicine in which people are treated to cure disease. Which of these diseases can be dealt with by preventive medicine, which can be cured by treatment with drugs and which can neither be prevented nor cured: cholera, cancer, diabetes, heart disease, AIDS, influenza, measles, tuberculosis (TB)?

20 Make summaries of the information in Chapters 2 and 3 about the bacterial disease TB and the viral disease influenza. Include in each summary the causes of the disease, the symptoms, methods of prevention, possible treatments and factors which make the disease hard to control.

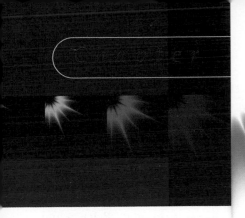

4

Health risks

The issues

People are living longer now it is known how to prevent or cure many of the infectious diseases which used to kill large numbers even in richer industrialised countries. As a result the main threats to health now, in countries like the UK, are diseases such as cancer and heart disease. Many factors can contribute to the development of these diseases including the genetic make-up of individuals, their diet, environmental pollutants they encounter, the work they do and the choices they make in how they live.

The media, school textbooks, health magazines and leaflets in doctors' waiting rooms all offer advice telling people how to live their lives in ways which will help them to cut down on the risks to their health. Often the link between cause and effect is uncertain or remote. Sometimes the advice is contradictory. It can be hard to take seriously today the advice not to smoke or to avoid fatty foods in order to limit the risk of lung cancer or heart disease tomorrow.

What this tells us about science and society

Experts and the general public react very differently to estimates of risk. Often the measures used to quantify risk have little meaning to individuals. People respond differently to situations where the risk is small but the consequences are very serious, and those where the risk is greater but the consequences are less severe. The way people respond to risks depends much more on whether or not the risk is imposed or taken voluntarily.

Many situations of importance to us are so complex that we cannot (yet) explain what causes different outcomes. Often the incidence of the condition is low – only a few individuals in a large population are affected. The best we can do is look for correlations between specific factors and outcomes. Many illnesses are influenced by several factors and it is difficult to show convincingly that a particular factor has an effect especially when the influence of the factor is small. Since there is always variation between any two population samples, it is very hard to gather evidence to show that the factor really does make a difference. It is even harder to show that a particular factor does not have any effect because two samples will always differ to some extent.

These uncertainties mean that people (including scientists) may interpret the evidence in quite different ways. Their backgrounds, personalities and interests can influence their interpretations. We need to take this into account in evaluating what they say.

Figure 4.1

Two-fifths of deaths among young men aged between 16 and 24 are from accidents.

Risk

Science and technology have done much to make our lives less risky. People in the industrialised countries live, on average, longer, healthier lives than in the past. But there are also growing concerns that science and technology may introduce new health risks – from the chemicals used in agriculture and food processing, from industrial wastes and from radiation, for example.

This makes it important to be able to assess how big different risks really are, so that we can make more rational decisions about whether to accept them, or try to reduce or eliminate them.

What do we mean by risk?

Life is uncertain. We might catch a cold in the next month, or we might not. The car journey we make this weekend might go uneventfully – or we might have an accident (Figure 4.1). There is a definite risk of these events happening. It is impossible to predict what will happen to an individual but we can calculate the probability of the event occurring by looking at larger groups of people. To measure the size of a risk, scientists look at the probability of the event occurring. So, for example, in a town of 6000 people, the evidence might show that 150 people caught influenza during one winter. Their risk of catching it that winter was therefore 1:40. In the same period, it might be that 600 people in the town caught a cold. The risk of this would then be 1:10 – rather higher than the risk of catching influenza.

Risk estimates like these are never exact. One reason is that someone might be diagnosed as having influenza when they only have a heavy cold. For this reason, the risk of death from different causes is easier to measure – as the outcome can be detected with certainty. Figure 4.2 shows the risk of a person in the UK dying in any one year from a number of different causes.

All statements about risk are estimates. Often it is difficult to measure the size of a risk, either because the outcome is hard to detect with certainty, or because the risk is low and only a few people are affected, even if data is collected from a large population.

The size of a risk may be very different for different groups within a population. For example, the risk of death from influenza in Figure 4.2 is an average for the whole population. The risk for people over 65 is much higher than 1 in 5000. The same is probably true about the risk of accidents in the home. This may be important in comparing risks.

Figure 4.2 Risk of an individual in the UK dying in any one year from various causes.	
Smoking 10 cigarettes a day	1 in 200
Influenza	1 in 5 000
Road accident	1 in 8 000
Accident at home	1 in 25 000
Radiation (for a worker in the nuclear industry)	1 in 50 000
Murder	1 in 100 000
Train accident	1 in 500 000
Lightning	1 in 10 million

Different ways of expressing a risk

Sometimes a risk is expressed in terms of the number of people likely to experience the event in question. For example, the population of the UK is 60 million, so the numbers in Figure 4.2 suggest that around 60 people will be killed by lightning over the next 10 years. This might seem higher to some people than a risk of 1 in 10 million. After the accident at the nuclear power station in Chernobyl in 1986, a spokesman for the National Radiological Protection Board said in a radio interview that the accident would probably result in around 30 cancer deaths over the next 50 years. This caused a considerable public reaction – whereas an equivalent statement a few days earlier that the accident would only raise the risk to an individual by less than 1 in 10 million passed without comment.

The size of risk may also depend on exposure to the hazard. For example the risk of dying in a road accident is greater for an individual who drives 20 000 miles a year than for one who drives 2000 miles a year. For this reason the death rate among car drivers is often expressed in relation to distance travelled. In Britain in 1996 it was 3 deaths per 1000 million kilometres travelled.

How people react to risks

People often do not know the actual risk for many of the activities they choose. They have to use their own perception of risk to decide whether it is safe to do something. However, research shows that people are not very good at judging risk from experiences.

People tend to overestimate the risk of unfamiliar or rare events, and underestimate the risks of familiar or common ones. If there has recently been an accident or a health-scare story in the news, people are likely to see the risk of this as larger than it really is.

Sometimes people are willing to accept the risk of an activity because of its benefits. So concerns about possible risks from mobile phones have not slowed the rapid growth in their use. And people still engage in 'dangerous' sports, despite the risks – because they enjoy them (Figure 4.4).

People are more willing to accept a risk from something they choose to do voluntarily, as compared to something they have no option about. So

Figure 4.3

Lightning striking the desert at night near Tucson in Arizona, USA. The risk of being killed by lightning is very low.

Figure 4.4

Hang gliding over snow and ice.

Figure 4.5

A smoking chimney near Pontypool in South Wales.

Figure 4.6

Hartlepool nuclear power station seen from the nearby town of Seaton Carew in the North East of England.

a person might willingly ride a powerful motorcycle, but would be horrified about a proposal to build a nuclear power station nearby.

People are more willing to accept a risk that ceases when you stop the activity in question – and more wary about risks that will continue for a long time. So people are concerned about the effects of chemical pollution, or radioactive materials, because the risk these pose may persist for a very long time (Figure 4.5). Even though hang-gliding might be much more dangerous, it ceases to have any risk once you stop doing it.

The hardest kind of risk to deal with is one which is very small – but with disastrous consequences if it does occur. The risk of a nuclear power station exploding is an example (Figure 4.6). The risk is extremely small. But if it were to happen, the consequences would be enormous. It is much harder to decide whether such risks are acceptable than ones where the risk is higher but the consequences less severe.

Risk factors

There are often stories in the newspapers and on television that something is believed to pose a risk to health, or to improve our health. For instance, living under high-voltage power lines has been claimed to increase the risk of some types of cancer. On the positive side, it has been claimed that drinking a glass of red wine a day reduces the risk of heart disease. Claims like these, however, are often contested. The findings are questioned. So what is involved in establishing that there is a link between a factor and an outcome – or in showing that there is no connection? 🌀 *Ref to Chapter 14*

Isolating one factor

Most things are influenced by several factors, not just one. This makes it difficult to spot a factor that has an important effect. If there is a suspicion that one particular factor might be important, then experts can collect evidence by comparing two situations: one in which the factor is present, and one where it is absent while keeping the same all the other factors that might be important. This is called a **cohort study**. Studies of this kind compare two groups of people (two cohorts): one group which has been exposed to the factor of interest, and another group which has not. This second group is the **control** group. The idea is essentially the same as the 'fair test'.

Setting up a cohort study is seldom easy. First, it is often not at all obvious which other factors do have an effect; so all that can be done is to identify all the likely ones and keep them the same. Secondly, if there are a lot of other factors to keep the same, it makes it difficult to find enough people for two groups of a reasonable size.

Detecting (or eliminating) small effects

If a factor causes a major effect, it is easy to spot – and people quickly recognise that it is important. But often a factor causes, at most, a very small effect. For example, worries about the effect of low levels of ionising radiation, or about eating beef, or about possible health effects of pesticides used in agriculture are like this. The number of cases is small. So if a study compares a group that is exposed to the factor with a control group, the numbers of individuals affected will be small in both groups. And if the numbers are small it is hard to be sure if any difference is real – or just due to random variation.

One way to tackle this is to use large samples. The bigger the samples, the more cases there will be in both groups – and the bigger any difference will be. The larger the difference, the less likely it is that it is just due to chance. But this may not solve all the problems. For instance, imagine that a study collects data about 10 000 people who have not been exposed to a certain factor, and finds one person with the illness. The study also gathers information from 10 000 who have been exposed and finds three cases. This is three times as many, but it is still just three cases. Maybe the extra two are just chance.

Another option is to use a **case-control study**. Again this compares two groups, but this time the groups are people who have the illness in question, and a similar control sample who do not. Then the people making the study look to see which people in the two groups have been exposed to the factor in question. This is like working backwards from the outcome to the factor – rather than vice versa. It has the advantage that it guarantees there are enough cases of the illness in the data set to reveal a pattern if there really is one. But it is harder to

Discussion points

Which are the big, voluntary risks in your life which you could do a great deal to control if you wanted to do so? What are the small, involuntary risks which you worry about but cannot do anything to change?

Contrary to what most people believe there are far more accidents, proportionally, in agriculture through carelessness than in industry from hazardous chemicals. Why do people respond in different ways to the accidents on farms and to the harmful effects of industrial chemicals?

Why are some people very concerned about the small and uncertain risks of eating genetically modified (GM) foods while willingly accepting the much larger and well-known risks of smoking or travelling by car?

Questions

1 How would you choose members of two groups for the following investigations? What would you want to keep the same in both groups?

a To find out whether taking more than 3 g of vitamin C per day reduces the risk of catching cold.

b To find out whether drinking coffee in the evening results in poorer sleep.

c To find out whether having a domestic pet reduces stress.

2 Toss the same coin 20 times and count the number of heads. Now do the same with another coin. The number of heads is likely to be different. Does this show that one coin is different from the other, in terms of its 'heads frequency'? Why not?

Question

3

a Imagine that you have two pennies, a normal one and one with a small bias to turn up heads. What problems would you have getting good evidence that one was biased?

b Now imagine that you have two normal pennies. How would you collect evidence to show that they were not different? Why would this be harder than showing that one penny was biased?

avoid bias. The people making the study are more likely to 'see' what they expect to see – even if it is not really there.

Interpreting information about risk factors

Sometimes claims about risk factors say that the factor increases the risk by so many per cent, or by so many times. It is, however, also important to know how many actual cases were observed. Figure 4.7 shows some data on two different risk factor claims.

Figure 4.7 Cases found in exposed and control groups of 10 000.

	Number of cases in samples of 10 000			
	Exposed to factor	Control	Risk increased by (%)	Risk increased (no. of times)
Claim 1	3	1	300	3
Claim 2	260	200	30	1.3

Although the percentage increase is much bigger in the case of claim 1, many people might be more convinced by claim 2, because the number of cases has gone up from 200 to 260. This seems unlikely to be just random variation, whereas an increase from 1 to 3 might be.

Convincing people there is no effect

Industries who want to persuade people that their activities pose no significant risk to health, or official regulatory bodies who want to establish that something does not pose a risk, have a particularly difficult task (Figure 4.8). There are always likely to be some differences between any two groups. So it is very hard to persuade people that a factor has no effect. This involves 'proving a negative'.

Epidemiology

The work of John Snow set the scene for the birth of epidemiology which is the study of the patterns of incidence of a disease. Many diseases, including cancers and heart disease do not have simple causes. Several variables are involved including dietary, environmental and genetic factors. Epidemiology is the branch of medical research which unravels the causes of disease by gathering data from large samples to test hypotheses suggesting connections between the incidence of the disease and aspects of lifestyle. *Ref to pages 4-6*

There are two main aspects of disease which are considered in epidemiological studies. One is the **morbidity** of the disease –

Figure 4.8

Advertisement from the nuclear power industry with statements about radiation.

how many people are actually made ill. This is not always as easy as it sounds. Morbidity is easy to record when a disease is easy to diagnose and causes clear-cut symptoms, but not all diseases are so easy to pinpoint.

The other aspect of a disease which can be measured is the **mortality** – the number of people who die of the disease. While this is easier to measure than morbidity, it is only useful in diseases serious enough to cause death. When these figures are used in association with social and biological factors they can lead to the discovery of the cause of a disease.

The results of epidemiological studies make it possible for doctors to identify risk factors for specific diseases. A risk factor is linked with a greater than average probability of developing that disease. Scientists are careful to distinguish between a 'risk factor' and the 'cause' of a disease. Epidemiological studies, however, often provide the first clue in the hunt for the causes of disease.

Health risks from smoking

In the first half of the twentieth century smoking was freely promoted and often glamorised (Figure 4.9).

Studies in the 1950s by epidemiologists such as Richard Doll changed attitudes. In 1971, for example, The Royal College of Physicians issued a statement:

> Premature death and disabling illnesses caused by cigarette smoking have now reached epidemic proportions and present the most challenging of all opportunities for preventive medicine...

Despite the evidence, however, large numbers of people still smoke, though fewer than at the time of the 1971 Royal College's statement. This despite the sustained campaign against cigarette smoking which has included initiatives to educate the public about the dangers of smoking, curbs on advertising tobacco products, a ban on smoking in many public places, and prominent health warnings on cigarette packets (Figure 4.10). The evidence that smoking causes ill health is now strong enough for people suffering ill health through smoking to win legal actions against cigarette manufacturers. Even the companies themselves now admit that smoking adversely affects human health.

The evidence

Research into the links between smoking and disease was prompted by the sort of data shown in Figure 4.11.

Notice that deaths from lung cancer increased sharply from the early 1900s when deaths from other forms of lung disease such as tuberculosis were falling. At first, scientists thought there might be a correlation between the increase in deaths from lung cancer and increasing air pollution because of the growing number of motor vehicles on the road. Cars were invented at the end of the nineteenth century and to make the link seemed reasonable. However, the research made little progress.

Questions

4 What was John Snow's hypothesis? What factors did he investigate? What data did he collect to test his hypothesis?

5 Why can it be useful to doctors and the public to identify risk factors for a disease even if it then takes a long time before the causes of the disease are understood?

Figure 4.9

A cigarette advertisement in 1934.

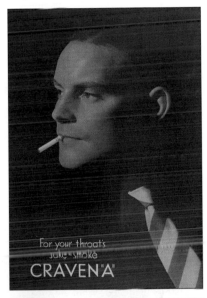

Figure 4.10

Health warning on a cigarette packet in 2000.

Question

6 What do these terms mean in the statement from the Royal College: 'premature death', 'disabling illnesses' and 'epidemic proportions'?

Figure 4.11

Deaths from lung disease in England and Wales, 1920–1960.

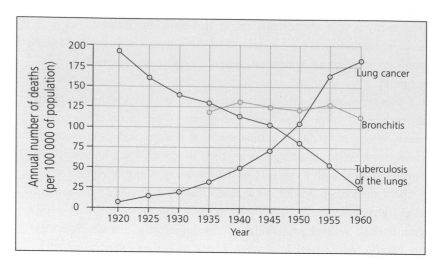

Questions

7 In what ways is a retrospective study similar to a case-control study (page 47)? What are the differences between the two types of study?

The habit of smoking cigarettes became widespread at about the same time as the appearance of the car, and attention turned from vehicle exhaust fumes to investigating the possible correlation between lung cancer and smoking.

Retrospective studies in the 1950s showed that over 90% of patients with lung cancer were tobacco smokers. Other studies have since shown that smokers are more likely than non-smokers to die of other respiratory diseases and heart disease.

Although it was difficult to prove the link beyond a shadow of doubt, when doctors saw the early evidence many of them gave up smoking cigarettes. Their swing away from smoking, in itself provided further valuable evidence of the dangers of cigarettes. Figure 4.12 shows that deaths from lung cancer among doctors went down compared with other people who did not understand as quickly the significance of the results from early research into smoking related diseases.

Other studies helped confirm the relationship between the risk of dying from lung cancer and the number of cigarettes smoked - the more cigarettes smoked, the greater the risk (Figure 4.13).

Since 1971, cigarette smoking among men and women in the UK has dropped. However, it is increasing among children and teenage girls in particular.

Figure 4.12

Comparing the death rates from lung cancer in male doctors and all men in England and Wales.

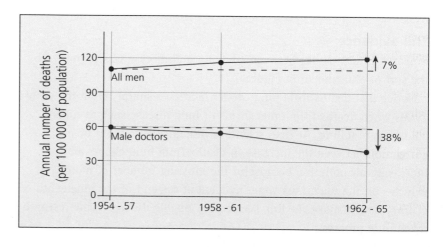

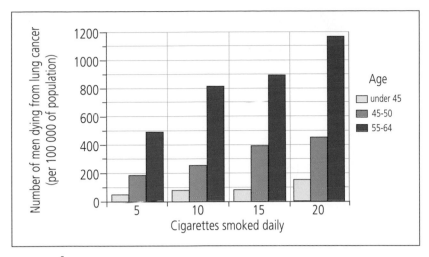

Figure 4.13

The relationship between the risk of men dying from lung cancer and the number of cigarettes smoked daily.

Smoking and lung disease

Even though the evidence suggested a strong correlation between smoking and lung cancer, scientists needed to provide a plausible explanation of why tobacco smoke could cause the disease. One approach was to analyse the chemicals in tobacco and tobacco smoke.

Cigarette smoke contains a mixture of nicotine and carbon monoxide and tar. All three have damaging effects on health. Nicotine is the drug which makes smoking a habit which is hard to kick because it causes addiction. It is a powerful drug which acts as a stimulant to the nervous system, increasing the heart rate and raising the blood pressure. Carbon monoxide is a toxic gas which reduces the amount of oxygen which the blood can carry. *Ref to Chapter 11*

Tar is a sticky brown mixture of some 4 000 different chemicals that accumulates in the lungs. About 60 of the chemicals in tar are suspected or known carcinogens. The theory is that these carcinogens damage the genes which control cell division.

In a healthy body, the rate at which cells divide to produce new cells is kept under very strict control so that it exactly matches the requirements for growth and for replacing cells that have been lost through wear, tear and injury. When a person has cancer, some cells are somehow able to escape from this control and multiply to produce a growth, or tumour.

Not only do the cells in the tumour make no contribution to the body, but they also take up space and get in the way of the activity of normal cells in the organs affected, eventually causing death. The ability of their cells to spread around the body and invade other tissues (metastasis) is one of the characteristics of cancer.

Carcinogens in cigarette smoke have been shown to cause mutations in the genes which control cell division. The mutations make the genes increase their activity so that they stimulate cell division. These mutated genes are called oncogenes and their effect is to cause cell division to

Questions

8 Why was it plausible that there might be a connection between the growth in motor traffic and the rise in lung cancer? Does Figure 4.11 offer any evidence for or against the possibility of such a link?

9 Why did it take a long time before people began to recognise the dangers of smoking?

10 Why were doctors among the first to respond to the evidence linking smoking to lung cancer?

11 What generalisations can you make based on the information in Figure 4.13?

12 Smoking remains as one of the greatest causes of preventable death worldwide. Each year in Britain about 120 000 people die early because they smoke . If this is the case, why do so many people smoke cigarettes?

13 How do you account for the fact that people aged 24 to 34 are most likely to smoke – around two-fifths of men and a third of women in this age group smoke compared to less than a fifth of adults aged 60 and over.

Key terms

Chromosome – a package of genes in the nucleus of living cells which consists of a tightly coiled length of DNA.

 Ref to Figure 7.2 on page 88

Gene – a segment of a DNA molecule in a chromosome which carries the code for making part of a living cell.

Carcinogen – any agent which causes cancer.

Mutation – a change in the genetic material of the cell which may be caused by chemicals or radiation. Mutations in body cells may lead, in time, to cancer, but they are not inherited. Mutations in sex cells can be passed onto the next generation; they are often harmful but may be beneficial.

 Ref to Figure 6.2 on page 76

Figure 4.14

Stages in normal cell division of a body cell to produce two new body cells. Each new cell has the same number of chromosomes as the original cell.

a Chromosomes become visible in nucleus.

b Chromosomes are distinct.

chromosome

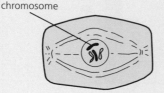

c Nuclear membrane breaks down: chromosomes (now formed of two chromatids) line up at the equator of the cell.

d The 2 daughter chromatids of each chromosome separate and move apart to opposite poles becoming daughter chromosomes.

e Daughter chromosomes clumped near poles.

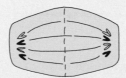

f Daughter chromosomes disappear; nuclear membrane reappears; cell has divided.

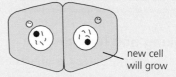

new cell will grow

When do cells divide?

Everyone starts life as a single cell when a sperm fertilises an egg. The new cell divides to form two cells, divides again and so on until it eventually ends up as a large multi-cellular person. For the first few divisions, the cells are all the same. Before long, however, the different types of cells begin to develop: nerve cells, muscle cells, blood cells and so on. Groups of specialised cells make up the tissues of human bodies. The various tissues are arranged to form our organs – heart, liver, kidneys, etc.

Even when people are fully grown, some cells need to be able to divide and produce new cells. By doing so they compensate for the cells that are lost through wear and tear. Cell division can also repair damage due to accidents or replace cells which only live for a short time such as red blood cells.

How do cells divide?

Consider a cell that has just been formed as a result of another cell dividing. Before the new cell can divide it must grow, duplicate all its complex chemical machinery and also make copies of the chromosomes in its nucleus. The chromosomes carry the information that controls the development and functioning of the cell. The chromosomes are mainly DNA. Sections of the long DNA molecules that provide the coded information on the chromosomes are known as genes. All cells have the same genetic information in their DNA but the information that is acted on in a particular cell depends on its local environment in the body. There is a control system which decides that in a specialised cell particular genes are 'expressed' but not others.

Questions

14 When doctors talk about cancer they may use these words: tumour, benign, malignant. What do these terms mean?

15 Identify and list some of the differences between normal body cells and cancer cells.

16 Why is it difficult to establish whether or not a particular chemical or mixture of chemicals causes cancer (is carcinogenic)?

run out of control. It is as if the chemical signal for cell division becomes jammed in the 'on' position. As a result, cells proliferate and form a cancerous tumour. This mechanism probably accounts for 25% of cases of lung cancer.

Tobacco carcinogens may also lead to mutations of other genes which inhibit cell division and suppress the formation of tumours. Their loss or inactivation by mutation contributes to the loss of control over cell division and subsequent development of cancers - especially lung cancer.

Explaining the onset of cancer is, however, difficult because there is usually a delay which may be as long as 10 to 20 years between exposure to the harmful agent and the onset of cancer. The sharp rise in lung cancer among men, for example, came 20 years after a similar rise in cigarette smoking. For women the same pattern was repeated a generation later. This delay made it more difficult to establish a link between smoking and the disease.

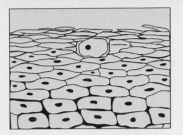

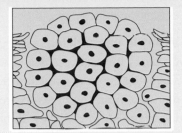

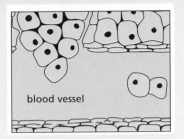

blood vessel

(a) Cells are dividing all the time. Usually the system works well. Occasionally the control programme breaks down in a cell, causing it to receive or give the wrong instruction.

(b) When its control programme breaks down, the cell becomes abnormal, and starts dividing in an uncontrolled way. These abnormal cells do not work properly as part of the organ or tissue where they began. As they multiply, abnormal cells take up more and more room. It may invade nearby parts of the body and prevent them from working properly.

(c) Cancer cells can break away and travel through the bloodstream, to other parts of the body, where they may form new colonies of abnormal cells. The new colonies are called secondary growths or metastases. Left unchecked, they too can prevent the organs where they're growing from working properly.

Nevertheless, it is now clear that there is a strong association between lung cancer and smoking. Scientists have their theories of the way in which chemicals in tobacco smoke can cause the disease. The evidence shows that giving up smoking reduces the chance of developing the disease but it is also the case that some who smoke heavily never develop lung cancer. So this is not a straightforward example of cause and effect.

Heart disease

The growth in the number of people dying of lung cancer followed the increase in the habit of smoking at a time when people were living longer. Heart disease is another condition which has become more significant in countries such as the UK since the discovery of ways to prevent or cure infectious diseases (Figure 4.16). In 1976, another report from the Royal College of Physicians attributed heart disease to lifestyle, smoking, fats in the blood, obesity and stress. This confirmed that many factors are involved in the development of heart disease. Genetic and environmental factors also seem to be involved.

Question

17

a What generalisations can you make based on the information in Figure 4.16?

b What explanations can you suggest to account for the generalisations?

Figure 4.16

Figures to compare the death rates from various diseases in developed and developing countries (1993). (Figures in brackets refer to the actual numbers of deaths in thousands.)

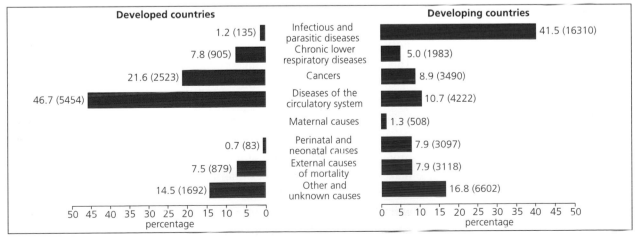

Developed countries		Developing countries
1.2 (135)	Infectious and parasitic diseases	41.5 (16310)
7.8 (905)	Chronic lower respiratory diseases	5.0 (1983)
21.6 (2523)	Cancers	8.9 (3490)
46.7 (5454)	Diseases of the circulatory system	10.7 (4222)
	Maternal causes	1.3 (508)
0.7 (83)	Perinatal and neonatal causes	7.9 (3097)
7.5 (879)	External causes of mortality	7.9 (3118)
14.5 (1692)	Other and unknown causes	16.8 (6602)

Whereas cholera is the result of infection by a specific bacterium and can be prevented by good hygiene, clean water and vaccination; heart disease is multifactorial and there are no specific ways of preventing or curing the disease.

What is heart disease?

Coronary heart disease is a disease in one or more of the main arteries supplying the heart muscle with blood. Like the other organs in the body, the heart too needs oxygen and nutrients. It may seem odd that the heart, which is filled with blood, needs its own blood supply. However, the walls of the heart are so thick that the oxygen and nutrients in blood cannot simply reach all of the heart muscle just by diffusion from the blood inside. Instead the coronary arteries, running over the heart, transport blood to the surface of the heart and to the tissue deep within its walls (Figure 4.17).

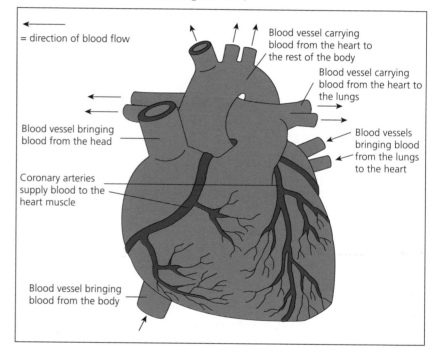

Figure 4.17

The coronary arteries pass over and penetrate the wall of the heart. They are called the coronary arteries because they circle the top of the heart like a crown.

The smooth inner wall of healthy blood vessels allows blood to flow through them easily. In coronary heart disease the smooth, glistening lining of the arteries becomes roughened and thickened mainly due to fatty deposits (atheroma) and hardened (sclerotic) by mineralisation. This leads to a condition which doctors call atherosclerosis.

In due course the fatty deposits may narrow the arteries so much that insufficient blood reaches the tissue beyond the constriction (Figure 4.18). If the coronary arteries are narrowed the first signs of trouble may be breathlessness and a cramp-like pain in the chest, brought on by quick walking, anger, excitement or anything else that makes the heart work harder than usual. Doctors call the pain angina. The pain is the heart's response to being starved of oxygen. People live with some types of angina for years, but other types get worse and may later result in a heart attack.

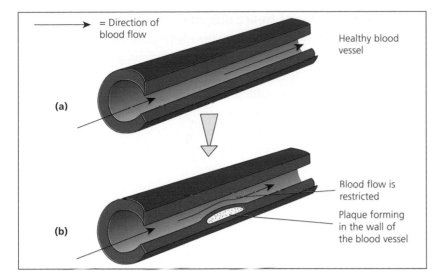

Figure 4.18

(a) Section through a healthy artery,
(b) Fatty material building up inside an artery.

= Direction of blood flow

Healthy blood vessel

(a)

Blood flow is restricted

Plaque forming in the wall of the blood vessel

(b)

A person suffers a heart attack when the reduction in blood supply beyond an obstruction in the coronary arteries is so severe as to interrupt the blood supply to the muscles. The ill person may feel sick and faint, and pain usually grips the chest, spreading to the arms, neck and jaw. Other signs of heart attack are sweating, breathlessness and a pale skin because blood is not reaching the body's surface.

When oxygen and nutrients cannot reach the heart muscle the tissue is damaged and may die. Doctors call the blockage a thrombosis - hence 'coronary thrombosis', a term often used to refer to a heart attack. Another common medical term for a heart attack is myocardial infarction. 'Myocardial' refers to the heart muscle and 'infarction' to the death of the muscle cells.

A severe heart attack may start a rhythm disturbance of the muscles in the ventricles of the heart, leading to cardiac arrest. The heart muscle is so disturbed that the heart cannot pump any blood. The person becomes unconscious, and the pulse and breathing stop. In such cases, it is essential to get the heart pumping again within a few minutes otherwise the person dies.

Question

18 Why may people with narrowed coronary arteries suffer from angina?

Figure 4.19

Heart attack: a plaque in the wall of a coronary artery ruptures and a clot forms.

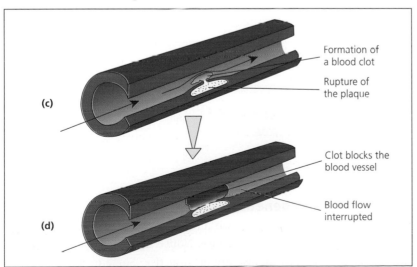

Formation of a blood clot

Rupture of the plaque

(c)

Clot blocks the blood vessel

Blood flow interrupted

(d)

In a **prospective study** the researchers recruit a number of healthy people and ask them to report aspects of their lifestyle, to give samples of their blood at intervals and to report any illnesses they suffer during the study period. They follow up the participants for a number of years and compare the diet and biological condition of those who develop diseases such as heart disease or stay healthy. Studies of this kind are expensive because large numbers of people have to be recruited and monitored over long periods so that the numbers contracting a particular disease are statistically significant.

The epidemiology of heart disease

Epidemiology has helped to unravel the likely causes of heart disease by studying patterns of disease in large populations and relating these patterns to aspects of lifestyle. Comparing groups of people who have high rates of heart disease with groups that have low rates has helped to identify some of the causes of heart disease.

Two prospective studies – the Seven Countries Study and the Framingham Study – have provided much of the evidence identifying the major risk factors for heart disease.

In the Seven Countries Study groups of men from different countries including the United States, Finland, Greece and Japan were examined and then followed up for 10 years. The study supported the hypothesis that the level of cholesterol in the blood was a risk factor. Men in the United States and Finland with high levels of blood cholesterol were significantly at greater risk from heart disease than men from Greece and Japan who enjoyed low cholesterol levels and correspondingly less heart disease.

In the Framingham Study, 5000 men and women were examined in the late 1940s and followed up for more than 20 years. The relationship between the level of blood cholesterol and the risk of heart disease confirmed the results of the Seven Countries Study.

Even so, it is important to recognise that there is no single cause of heart disease. For example, heart disease is relatively rare in Japan but common in the UK and Finland as Figure 4.20 shows. Quite why remains a matter of investigation and debate.

Figure 4.20

Number of deaths from heart disease (35–74 year olds) per 100 000 population for different countries.

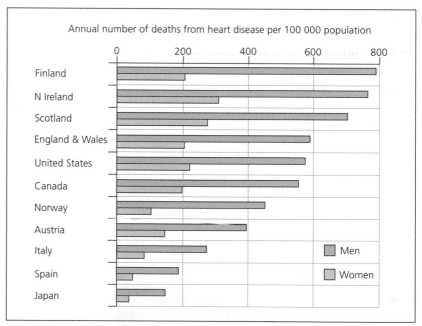

Annual number of deaths from heart disease per 100 000 population

Questions

19 In Figure 4.20:

a what is the ratio of the number of deaths from heart disease for men in Finland to the number of deaths from heart disease for men in Japan?

b what is the highest ratio of the number of deaths from heart disease for men to the number of deaths from heart disease for women? In which country is this observed?

20 Having studied Figure 4.20 what questions would you want to ask an expert in the epidemiology of coronary heart disease?

Gender, age and genetics

Our gender, age and the genes we inherit are unavoidable risk factors. Figure 4.21 summarises the evidence. Notice that older men with a family history of heart disease are most at risk. Some or all of the

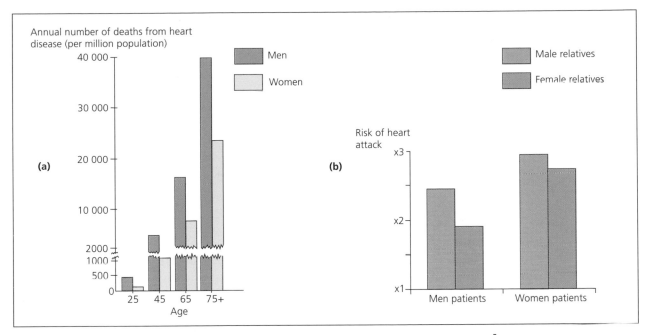

Annual number of deaths from heart disease (per million population)

(a)

Age

(b) Risk of heart attack

Men patients   Women patients

Men
Women

Male relatives
Female relatives

Figure 4.21

Unavoidable risk factors: (a) Deaths from heart disease according to age and sex, for England and Wales, (b) Relatives of people who have a heart attack are more likely to suffer one themselves. Notice that the relatives of women patients are more at risk then the relatives of men patients. Is the inheritance of risk from heart attack sex-linked?

unavoidable risk factors are a part of everyday life, but it does not mean that people cannot increase their chances of reaching a ripe old age without developing heart disease. Reducing exposure to the avoidable risk factors in our daily lives is the sensible strategy for maintaining a healthy heart.

Diet and levels of cholesterol

People who eat too much fat and sugar tend to put on weight. Excessively overweight people have a higher risk of heart disease, not only because the excess weight is a risk factor in itself (Figure 4.22) but also because overweight people are more likely to develop diabetes and high blood pressure.

Putting on weight is not the only problem of eating too much fatty food. The level in the blood of a substance called cholesterol may also increase. Cholesterol is a clear, oily liquid and is often linked with heart disease. It is part of animal cell membranes and a component of steroid hormones such as testosterone and oestrogen.

Large amounts of cholesterol are found in the deposits blocking blood vessels. Blocked blood vessels cause heart disease. Figure 4.23 shows that the more cholesterol there is in the blood, the greater the risk of heart disease. Eating food containing a lot of saturated fat seems to be the problem. Saturated fats raise the natural level of cholesterol in the blood.

However, the story is more complicated than appears at first sight. The liver makes cholesterol from the breakdown products of fats. As the intake of fat formed from saturated fatty acids increases, the liver makes less cholesterol, so adjusting blood cholesterol levels within normal range. It is only when the intake of saturated fat is more than the liver can adjust for, that high blood cholesterol levels result. Even so, not everyone with a diet rich in saturated fats will develop high blood cholesterol levels. Some people seem to be able to break

Question

21 In Figure 4.21 what do you notice about the risk of heart disease for the relatives of women patients compared with the relatives of men patients? What does this suggest?

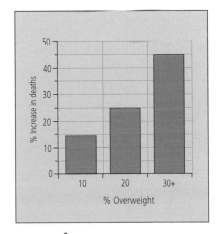

Figure 4.22

Increase in deaths from heart disease due to being overweight.

Figure 4.23

The relationship between blood cholesterol levels and risk of death from coronary heart disease.

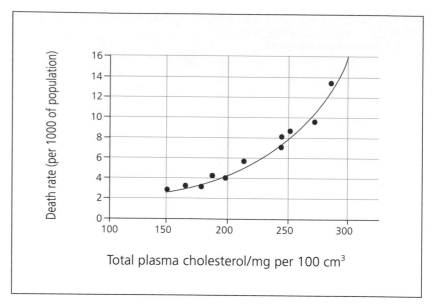

down cholesterol more effectively than others. Also, in some people, the liver produces less cholesterol. Blood cholesterol levels, therefore, need not have a dietary link.

Particular genetic characteristics cause the liver to produce too much cholesterol. Even if people with a genetic tendency to make too much cholesterol stick to a low cholesterol diet, they still generally have blood cholesterol levels higher than most other people.

Cholesterol is insoluble and transported in the blood bound to a carrier protein. The combination of cholesterol and the carrier is a lipoprotein. There are several types of lipoprotein, each one consisting of varying amounts of protein, fat and cholesterol. They are identified by density. Low density lipoproteins (LDL) and high density lipoproteins (HDL) are especially important in the development of heart disease. The evidence shows that raised levels of LDL increases the risk of heart disease, whereas raised levels of HDL lowers the risk of heart disease.

Figure 4.24

The consumption of saturated and unsaturated fat. Low and reduced fat spreads contain less saturated fat than butter, hard margarine and lard. In 1997 the consumption of butter was half that in 1984 while over the same period consumption of low fat spreads increased more than six times.

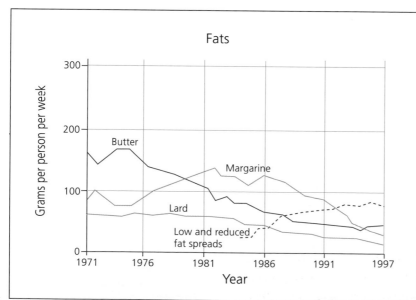

Measuring the proportions of LDL and HDL in the blood is a better risk indicator for heart disease than measuring levels of cholesterol alone.

Apart from reducing the intake of saturated fat, how else can dietary targets which reduce levels of blood cholesterol be achieved? Eating fatty fish seems to be beneficial. The evidence is based on the observation that Inuits who maintain a traditional diet and eat large quantities of fish, are relatively free of heart disease. It seems that fatty fish such as herring, tuna, trout and salmon contain substances that can help prevent the build up of the deposits that block blood vessels. Bran from whole oats also helps to reduce levels of blood cholesterol. It binds with cholesterol, partly blocking its absorption through the wall of the intestine. As little as two tablespoons of oat bran a day may reduce blood cholesterol levels significantly.

Overall the evidence suggests that total food energy intake should consist of no more than 30% as fat, and that the fat eaten should contain slightly more unsaturated fat than saturated fat.

Stress and heart disease

The link between stress and heart disease is less obvious than the links with diet, smoking and high blood pressure. However, worry, anxiety or frequent crises raise blood pressure and perhaps increase the risk of heart disease. Personality also plays its part. Some people are ambitious, competitive and seem to be always cramming as much as possible into the time available. They are **Type A** personalities. Their opposites are **Type Bs** who are more relaxed and easy-going, taking crises calmly in their stride. The evidence shows that Type A people have higher levels of stress hormones like adrenaline in their blood. The hormones raise blood pressure and cholesterol levels increasing the risk of heart disease.

A study of men who worked for the Western Electric Company, USA, showed that Type A people also tend to imagine threats from others and feel angry and hostile towards them. It rated the men for hostility and then followed them up 20 years later to find out how many had died from heart diseases. A study of medical students was followed up in a similar way. Figure 4.25 shows the results.

Questions

22 Summarise in 100 words with one or two diagrams the key points about diet, cholesterol in the blood and heart disease in a style suitable for a leaflet or poster addressed at patients waiting to see a doctor.

23 How do you account for the changing patterns in the consumption of solid fats from 1971 to 1997?

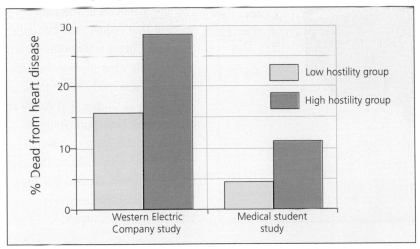

Figure 4.25

Hostility and heart disease.

Anger and hostility are stressful. If we try to accept that working conditions and the people we work with cannot always be changed and try to divert our stress into other channels, then effective stress management is more likely.

However, what is stressful to one person is not so to another. Some stress is a normal and healthy part of everyday life. It helps people keep alert and out of danger. Every time people cross the road, take

Questions

24 What difficulties are there in classifying people's personalities as Type A or Type B?

25 What assumptions do you have to make if you want to argue that Figure 4.25 supports the hypothesis that there is a connection between the incidence of heart disease and stress?

part in an argument or watch an exciting film their stress levels increase for a while. It is when stress levels stay high for months or years that trouble starts and the heart may suffer.

The effects of exercise

Ageing begins in the mid-twenties. This is ageing in its biological sense of loss of fitness rather than advancing years. Different aspects of fitness peak at different times and fitness of the heart is part of the picture. A fit heart pumps more blood than does a less fit heart. Regular exercise increases heart size and the volume of its chambers, and improves the efficiency of the contraction of heart muscle – factors which increase heart fitness. The combination of more efficient heart muscle and reduced oxygen demand by the heart at rest and during exercise makes life-threatening disturbances of the heart less likely.

Exercise may help make the heart more efficient but that does not mean that it cuts the risk of heart disease. The London Transport Study published in 1953, was one of the first investigations into the relationship between physical activity and heart disease. Figure 4.26 summarises the 2-year investigation. The conductors were more active than drivers, walking the length of the bus and up and down the stairs collecting passengers' fares. By comparison, drivers were sitting for most of the day. Analysis of the data showed that conductors not only had fewer heart attacks than drivers, but what disease they did have was less likely to be fatal. The different studies that followed the London Transport Study all help make the case: regular exercise not only reduces the chance of heart disease developing, but also helps recovery following a heart attack.

Figure 4.26

The London Transport Study.

The London Transport Study

31 000 central London male bus conductors and drivers studied

Not much exercise sitting down all day

Running up stairs keeps me fit

Drivers had twice as many fatal heart attacks as conductors

Conductors had 50% fewer heart attacks than drivers

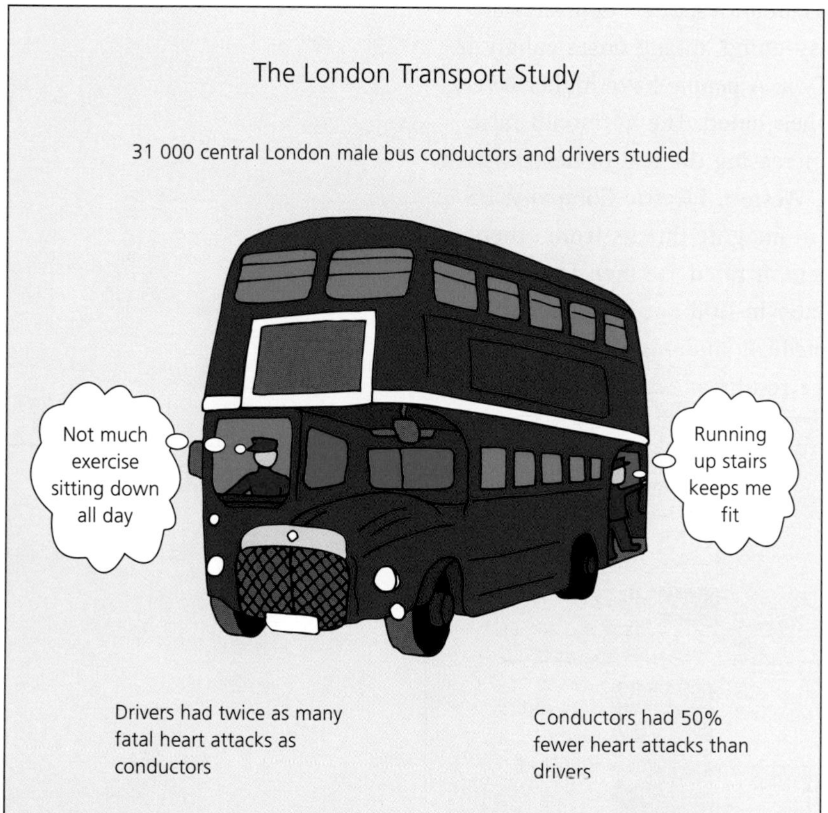

Adding up the risks

The unavoidable risks of heart disease set the risk rate. If, for example, members of a family are long-lived and have low blood pressure and low blood cholesterol, then a slight weight problem is unlikely to increase by very much their risk from heart disease. However, if the circumstances of their risk factors are the other way round, then they would be unwise to smoke cigarettes and should be careful of their weight.

The individual must know the risks and come to a balanced pattern for living. Figure 4.27 makes the message clear. It shows zero to three risk factors coming together in a man. Three risk factors together make him nearly three times more at risk from a heart attack than someone with fewer risk factors.

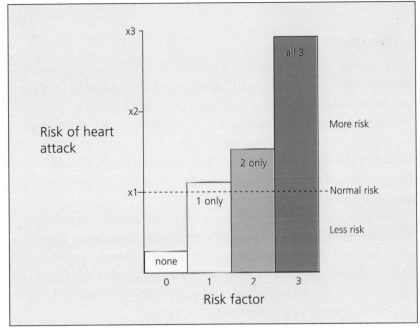

Figure 4.27

Effects of risk factors working together, increasing the risk of heart attack.

Questions

26 Do the results in Figure 4.26 clearly show that exercise helps to reduce the risk of a heart attack? What other differences between the lifestyles of drivers and conductors might be significant?

27 Why is it more difficult to identify the causes of a condition such as heart disease as compared to lung cancer or smoking?

Review Questions

28 Is it necessary to have an explanation for the way in which risk factors cause disease in order to benefit from the findings of epidemiological research?

29 Policies for lowering the rates of heart disease can involve:

- primary prevention which involves measures to encourage the public to change their lifestyles to cut the risk of heart disease, and

- secondary prevention to identify those at risk of heart disease or at an early stage of the development of the disease (for example, by screening for high blood pressure or levels of cholesterol in the blood and then prescribing treatment where necessary).

Compare primary and secondary prevention and discuss the advantages and disadvantages of the two approaches. Which should have a higher priority in health policy in the region where you live?

5

Alternatives
in medicine

The issues

The discovery of antibiotics meant that for the first time doctors could prescribe effective cures for infectious diseases. The control of infection also made possible surgery such as open heart operations, hip replacements and kidney transplants. In the same period other medical successes included a vaccination programme to eliminate smallpox and the use of insulin to treat diabetes. These triumphs meant that doctors, pharmacists and many members of the general public became hugely confident about science-based medicine.

Now people are realising that there are limits to high technology medicine based on drugs especially if they suffer from long-term illnesses which do not seem to respond to conventional treatments. More and more people are turning to alternative, traditional forms of therapy. Even doctors are prepared to regard some alternative therapies as legitimate types of complementary medicine.

This creates difficult choices for individuals who may find it hard to decide when to trust the local GP or hospital and when they should consider turning to one of the alternatives such as homeopathy or acupuncture.

What this tells us about science and society

Scientists value observations and measurements which can be repeated by different people in different places with the same results. So scientists are uneasy when faced with the results of complementary medicine because the results are often linked to the treatments offered by individual therapists working with patients who have particular faith in their treatments.

Scientists find it much easier to accept that a cure is a result of a particular treatment if they can understand how it works, or even imagine how it might conceivably work in terms of currently accepted basic ideas. Some types of complementary medicine seem to lie outside the world of conventional science. Homeopathic cures, for example, use such heavily diluted materials that it is hard to see how there is enough active ingredient present to have an effect.

So complementary medicine challenges science in its claims about cause and effect. However, scientists have been forced to take notice of the popularity of some alternative treatments and the successes of complementary therapies in treating conditions which conventional science-based medicine is unable to cure.

Figure 5.1

Dispensing homeopathic pills over a homeopathic textbook. Homeopathic remedies such as these defy the logic of scientists and have not undergone rigorous scientific testing before use – yet for centuries they have healed people.

What is health?

One of the commonest ways to greet someone almost anywhere in the world is to ask 'How are you?' – yet health is very difficult to define and measure. Perceptions of health also depend on the social context in which they occur. A slight headache on a school or college day may be enough to keep you in bed feeling unwell. The same symptom on a weekend or holiday might easily be ignored.

In the UK the main aims of official policy are to improve the health of the population as a whole by increasing the length of people's lives, by extending the number of years that people spend free from illness and by narrowing the health gap between those worst off in society and those that are better off.

An indicator which is often used to compare the health of nations is life expectancy. Since 1971 the life expectancy in the UK has been rising at a rate of about 2 years each decade for men and one and a half years for women so that by 1997, the average life expectancy for men was 74 years and for women 79 years. Recent research, however, suggests that while life expectancy is increasing, the expectancy for a healthy life has remained constant in the same period. On average men can expect 59 years of good health, while for women the figure is 62 years. So people suffer from some degree of illness or disability during the extra years of life.

Another indicator used to compare the health of populations is infant mortality. In the UK the infant mortality rate has halved since 1981. In 1997 the average rate was just under six deaths per thousand live births. This average figure hides the marked differences in infant mortality rate between the children born to families with fathers in professional occupations which is half the rate for infants in families to which the fathers are unskilled.

Figure 5.2 *Health problems reported by men and women by age, 1996–97.*

United Kingdom	Percentages				
	16–44	45–64	65–74	75 and over	All aged 16 and over
Males					
Pain or discomfort	18	39	52	56	32
Mobility	6	22	36	50	18
Anxiety or depression	12	19	20	19	15
Problems performing usual activities	5	16	21	27	12
Problems with self-care	1	6	8	14	5
Females					
Pain or discomfort	20	40	51	65	34
Mobility	6	21	37	60	19
Anxiety or depression	18	24	25	30	22
Problems performing usual activities	7	17	23	40	15
Problems with self-care	2	5	9	21	6

Questions

1 What are the advantages of 'life expectancy' as a measure of the health of a population? What are the disadvantages?

2

a What generalisations can you make about the health problems which people report in the survey results in Figure 5.2?

b What does Figure 5.2 suggest about people's views on 'good health'?

3 Suggest reasons to account for the differences in the health of people in different social classes in the UK. Why is it difficult, in prac-tice, to narrow the health gap?

Discussion points

Do you think that health is simply absence of physical illness or is there more to it than that? Explain your answer. How do your views compare with the views suggested by the results of the surveys reported in Figures 5.2 and 5.3?

The World Health Organisation (WHO) defines health as: 'a state of complete physical, mental and social well-being which is more than just the absence of disease'. Assess your own health using this definition. Are you healthy? If not, list the factors which stop you from fulfilling that definition of total health.

To what extent are you, your friends or your family prepared to put effort not just into getting better when ill but into taking positive steps to remain healthy in the first place? Is the attitude of your gener-ation towards health similar or different from your parents or from your grandparent's generation?

Do governments have the right to make people healthy by methods which strongly encourage them to alter their life style? Should there be financial incentives, for example, to encourage people to avoid smoking or excessive drinking?

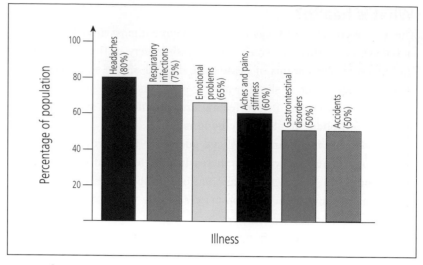

Figure 5.3

The percentage of adults suffering from minor illnesses in a single year in the UK.

The regular general household survey shows that there are people in all age groups whose lives are affected by illness or disability. Figure 5.2 lists the health problems reported by men and women in 1996–97. A common type of complaint reported in every age group for both sexes is conditions which affect the muscles or skeleton such as rheumatism and arthritis.

A survey to monitor the impact of health education in 1996 showed that people are aware of the connections between lifestyle and health. Over half of the men surveyed and three in five of the women thought that their relationships and their lifestyle were good influences on their health. The other positive influences reported were diet and exercise. In the same survey people were asked what they thought to be bad for their health with the results shown in Figure 5.4.

Health and medicine

The tasks of medicine are to comfort and to cure. Disease can be literally dis-ease and the comfort or removal of the stress may be all that is needed to bring about healing.

Some approaches to medicine emphasise comfort with an approach to therapy based on an overview of the person's mind, body and environ-ment. This is a holistic approach which sees the patient as a whole person and not as a biological machine consisting of interacting organs.

Science-based medicine has tended to put the emphasis on cures targeted at particular causes since the series of successes based on the germ theory of disease. As a result the National Health Service (NHS) in the UK is mainly an ill-health service which is good at intensive care in emergencies but much less good at encouraging good health.

Orthodox medicine expects high standards of evidence when judging alternatives such as acupuncture or homeopathy but can be less rigorous when assessing the effectiveness and safety of its own proce-dures. Despite its many successes, conventional medicine also includes

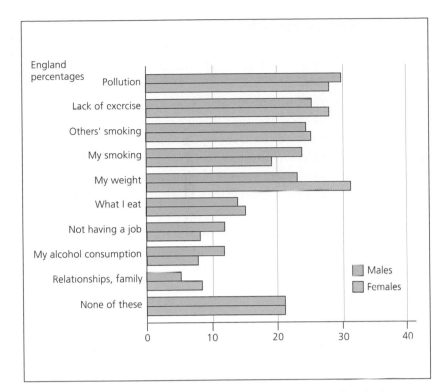

England percentages

Pollution
Lack of exercise
Others' smoking
My smoking
My weight
What I eat
Not having a job
My alcohol consumption
Relationships, family
None of these

☐ Males
☐ Females

0   10   20   30   40

Figure 5.4

What adults think is bad for their health as reported by the Health Education Monitoring Survey in 1996.

Discussion points

Which forms of health care now available in the UK place more emphasis on caring as a means of restoring health and harmony in people's lives?

Currently we invest heavily in a health service which puts the emphasis on curing through intensive treatments targeted as specific causes of diseases. Would it be more effective and more equitable to spend much more of the money on measures to prevent disease and maintain good health? What are the barriers to a change of policy along these lines?

Question

4 What harm can arise if people believe that there is a drug to cure everything?

untested and even dangerous treatments. Around 60% of orthodox medical procedures have never been subjected to rigorous trials.

Methods of treatment can become established over a number of years; they may not do too much harm but they may not do much good either. Results of a study published in September 1999 claimed that a standard treatment for severe shock and burns used for around 30 years had caused an estimated 1 200 deaths a year in the UK alone. Then in October 1999 came an even more shocking revelation. In the 1980s a heart drug called 'Lorcainide' caused the deaths of around 80 000 people in the USA. These deaths could have been avoided if negative results in some of the drug trials had been made public. But these were suppressed and only the positive results were published, with tragic effects for many of the patients subsequently prescribed the drug.

Orthodox medicine is the answer for many conditions and can often effect a complete cure. But there are many other conditions, often not life threatening but causing chronic, long-term discomfort, where orthodox medicine does not seem to have the answer. The orthodox solution is often seen to involve treatment with drugs, but these chemicals can have very unpleasant side effects. Back pain is the single biggest cause of missed time at work in the UK. Eczema, asthma, allergies, general fatigue and headaches do not always respond well to orthodox medicine. Some diseases have a strong psychological component; they are made worse by anxiety and may be relieved by placebos. Often the treatment offered merely masks and relieves the symptoms rather than effecting a cure. Increasingly people with these types of problems are turning to alternative therapies – the world of complementary medicine.

Questions

5 Why are so many medical procedures not based on rigorous clinical trials?

6 What are the pressures and difficulties which mean that pharmaceutical companies may not always pay proper attention to harmful side-effects of new drugs? What procedures or regulations might be put in place to ensure that dangerous side-effects are not ignored?

Complementary medicine

Complementary medicine is a term which covers an enormous range of different approaches to health care and healing which differ from orthodox medicine. It covers a wide range of therapies which include acupuncture, chiropractic, herbalism, homeopathy, reflexology and many more. Most of these therapies involve considerable consultation with the patient, but some can take place without the patient even being present. Many complementary therapies have been around for hundreds if not thousands of years, but some are relatively very new.

The common theme in complementary medicine is that the therapies aim to enhance the natural response of the body to illness and to allow for a natural healing process. This is in contrast to orthodox medicine which tends to involve direct intervention by drugs or surgery.

People turn to complementary therapies for a variety of reasons. If orthodox medicine has failed to cure an illness then an alternative may be sought. Some people do not like the idea of long term drug therapy and are looking for an alternative way to manage or cure their condition. Yet others simply want to take more control over their own health in what they perceive to be a more natural way.

Some complementary medicine involves ideas about the 'energy' of the body. The treatments claim to help to channel the natural energy of the body, or help the 'energy' to flow. It may involve the unblocking of channels for the energy to flow or the stimulation of ancient 'energy' lines. By making use of the natural energy of the body healing can take place – this is the model which much complementary therapy uses to explain its effects. This, however, is a major problem for orthodox medicine and medical scientists – the 'energy' which is so important in complementary therapies cannot as yet be measured in a scientific way, and the 'energy channels' do not seem to exist when a body is dissected. This means that the explanations for the way complementary therapies work does not fit with accepted explanations for body function, and this makes it more difficult for many scientists and doctors to accept that it could be effective.

The therapies can be divided into several groups:

- nutritional medicine, clinical ecology, herbalism and homeopathy claim to help the whole body to work better;
- acupuncture and healing claim to free up the flow of energy around the body;
- mind-body therapies which relax and focus the mind;
- osteopathy and chiropractic which make structural changes in the body.

A choice of therapies

People using orthodox medicine know exactly where to go if they are unwell – they visit their doctor. But when people use alternative therapies, they have to decide which one to choose. Different therapies

Discussion point

What experience do you and your friends, acquaintances and families have of complementary medicine? Which are the popular forms of therapy? What sorts of people look for alternatives to conventional medicine? How do they respond to treatment?

seem to work better in different conditions, but to add to the confusion some conditions can be helped by a variety of different therapies – it all comes down to personal choice.

One clear example is the case of allergies. Allergies have become increasingly common over the last 20 to 30 years, affecting people of all ages. They are caused by an overreaction of the immune system to a substance in the environment. When the immune system goes into over-drive it can cause symptoms which range from uncomfortable to life-threatening. Hay fever, asthma, perennial rhinitis (a constantly runny or stuffed up nose), and one type of eczema (red, inflamed, itchy and flaky skin) are all examples of illnesses caused by allergies (Figure 5.5).

When the immune system becomes over-sensitive the body itself comes under attack, which can lead to distressing symptoms like this eczema – and much worse.

The body may be sensitive to a whole range of things which can include grass pollen, fungal spores, animal hairs, dust mites and certain types of food or drug. The allergic response to some foods (peanuts are a well known example) can be very violent, with swelling of the lips, mouth and tongue, nettle-rash, vomiting and in really severe cases collapse and death as the swelling in the mouth blocks the airway and the heart fails. These violent allergies are terrifying, and sufferers usually carry adrenalin, a hormone which can keep their heart working and anti-histamines which reduce the reaction of the immune system and bring down the swelling if these symptoms start to develop. Conventional medicine provides life-saving solutions when the allergy is this violent.

However the majority of allergy sufferers have much milder symptoms, and for them orthodox medicine can offer only drugs to take on a long-term basis. Increasing numbers are looking for alternative ways to treat their allergies – and there is a wide choice open to them. Acupuncture (see below) offers help to sufferers of both asthma and eczema, as does homeopathy.

Homeopathy is based on the principles of 'like cures like' and 'less is more'. Very small quantities of whatever is causing an illness are given to stimulate the body's defence systems. The assumption is that the remedies used encourage the body's ability to heal itself. Treatments are based on very dilute solutions of healing herbs. The solutions are so dilute that the remedy is almost pure water.

Some chiropractors claim that chiropractic can restore balance and health in the body and thus enable it to overcome allergic illnesses. Chinese herbal medicine can be helpful for eczema and hay fever while nutritional therapy can help with asthma and food allergies (Figure 5.6). Clinical ecology is concerned with the environmental impact of air, water, food and drugs on people's health and also with environmental chemicals in our habitat; it too can help reduce both asthma and eczema.

There can be hazardous side effects from some alternative treatments too. Some herbal remedies contain powerful active ingredients which have to be used with care. A problem with unlicensed remedies is that

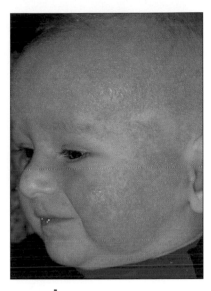

Figure 5.5

An eczema rash covering most of the skin on a baby's face.

Questions

7 Allergies are common. Make a list of everyone you know who suffers from some sort of allergic reaction – hayfever, asthma, eczema and so on. If possible note what causes the allergic response. Combine the results of your whole group and discover which types of allergic reaction and which allergy triggers seem to be the most common. Make a bar chart to display your results.

8 Why do you think people suffering from allergies turn to complementary therapies? Do you think they are wise to do so?

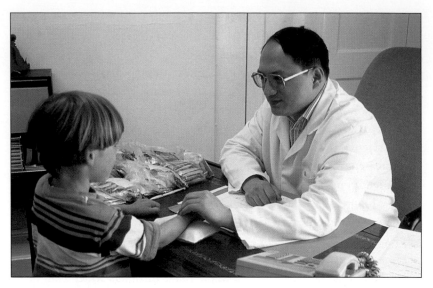

A Chinese herbal doctor taking the pulse of a young boy. On the desk are various herbal remedies.

the levels of active ingredients can vary from one preparation to another. An example is the preparations made from St John's Wort. New evidence suggests that St John's Wort treatments can interact with a range of drugs including oral contraceptives, some anti-convulsant drugs and drugs taken by HIV patients.

Does complementary medicine work?

Growing demand

Many people clearly believe that complementary medicine works. This is demonstrated by the growing number of people who are turning to alternatives to NHS treatment either for a cure for their complaints or to help them maintain a sense of well being. Twenty years ago there were about 13 500 practitioners of different types of complementary medicine in the UK. Now there are more than 40 000. In 1998 at least 15 million complementary medicine consultations took place in the UK. More than 40% of GP practices now offer some form of complementary therapy to their patients, and 70% of it is paid for by the NHS.

A survey carried out by *Which?*, a consumer magazine, found that 75% of the patients who tried complementary medicine felt that their condition had improved, and 83% felt that their general sense of well-being was better. So the general consensus of people who have tried complementary medicine is that, yes, it works.

One of the biggest problems with assessing complementary medicine is that it is quite difficult to find hard evidence for its effectiveness. The therapies tend to work on a number of very subtle levels, and often the close and confidential relationship with the therapist is important. A therapy may make a patient feel better almost immediately, even if physically measurable symptoms remain. Does this matter if the patient is now pain-free and feeling healthy? Because of difficulties such as these complementary therapists can be loathe to subject their work to scientific evaluation – they and their patients believe that it works, and people continue to come back – and pay for – continued treatment.

Two case studies

When Geoff was born one of his feet was turned inwards. Doctors decided there was nothing they could do until he was older. By the time he was 10 years old the twisting of his lower leg was so bad that Geoff could not take part in sport at school, and his parents had been told that the only option was to operate, breaking the leg bones and rotating them to bring the foot into alignment. Desperate to avoid such surgery if they could, Geoff's parents took him to a chiropractor recommended by some friends. A chiropractor has undergone a long training – in the USA this lasts 6 years and is on a par with medical training.

Chiropractic manipulation involves working on the joints between the vertebrae of the spine (Figure 5.7). After a few sessions both of Geoff's feet faced forwards, and his mother was almost in tears at sports day as she watched him come third in the 100 m sprint. Several years on, his foot is still facing the right way and he will certainly not be having conventional surgery.

When Ann entered her fourth pregnancy, she knew what to expect. For about 20 weeks in each of her earlier pregnancies she had been sick, vomiting violently up to 40 times a day and feeling desperately ill. With three small boys to look after the prospect of weeks of sickness and bed rest was almost unbearable – but she would not put her baby at risk by taking anti-nausea drugs. When the vomiting started she decided to try acupuncture. Needles were positioned in her wrists and below her knees and left for about 20 minutes. The treatment relieved the symptoms and stopped the vomiting – but only for 24 hours. After three sessions Ann's husband was taught how to position the needles so the treatment could be given daily – and cost nothing. The sickness was conquered, and the time intervals between the acupuncture treatment got longer and longer. From only ten weeks into the pregnancy the vomiting and nausea had completely stopped.

The problem for scientists and doctors is that they find many reasons not to accept anecdotal evidence, yet hard scientific evidence is in short supply. When studies have been carried out they have often been done by enthusiasts rather than scientists, so the methodology of the investigation is often open to question. For example, many studies have been carried out on very small groups of patients, or on self-selecting groups for whom the treatment has worked without reference to people who may have received little benefit. Evidence of this type simply strengthens the hand of those wishing to demonstrate that all alternative medicine is just so much hocus-pocus.

Research or recommendation?

In some areas of complementary medicine a substantial body of evidence is being built up by orthodox scientific testing. Dr George Lewith is a medical doctor who now lectures in complementary medicine at the medical school at Southampton. He and his team have set up a number of pieces of conventional research into the effective-

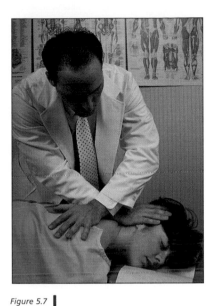

Figure 5.7

A chiropracter at work with a patient.

Discussion point

Why does anecdotal evidence like this have such a powerful influence on ordinary people? Why is the scientific community much less impressed with anecdotal evidence? What type of evidence do scientists need to see before they accept a new theory or treatment?

9 Why is it often impossible to carry out a placebo-controlled, double-blind trial (see pages 37–38) to test a form of complementary medicine?

10 Why do most members of the public take so little notice of scientific findings unless they are presented in the popular press?

Figure 5.8

A Chinese acupuncture chart with illustrations of internal organs. This illustration is based on the 'Bronze Man' of the Song dynasty (960–1279) and was produced in 1906. One reason for the disbelief of scientists in complementary therapies such as acupuncture is that the explanation for how it works does not fit in with the accepted scientific view of the physiology of the human body.

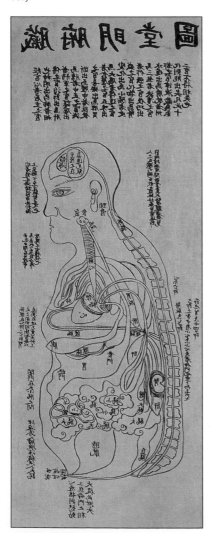

ness of acupuncture and other complementary techniques. Around the world other similar projects are underway and so gradually, in some areas of complementary medicine, scientifically acceptable evidence for their effectiveness is building up.

This is where private individuals often differ from the scientific community. Most people will not choose a particular type of complementary therapy because they have read an excellent and rigorous research paper showing that it is effective. They will go to someone recommended by a friend or a relative, someone who has had a similar problem solved. The choice of whether to use a chiropractor, a homeopath or an acupuncturist will be based not on empirical evidence as to which therapy is best but on which one worked best for someone they know.

Acupuncture

The Chinese tradition

Acupuncture is an ancient technique which originated in China where it has been practised for at least 2500 years. It involves inserting hair-thin needles into very specific points on the body to prevent or treat illness. Acupuncture is part of the Chinese holistic system of medicine, which views health as a constantly changing flow of energy, **chi** or **qi** (pronounced chee). Imbalances in this natural flow of energy are thought to cause disease, and the application of needles in acupuncture helps to restore the balance of the flow.

According to the principles of traditional Chinese medicine, chi flows along 14 primary meridians or channels through the body. Inserting needles into specific places along these channels (known as acupoints) strengthens the flow of chi or removes blockages. There are thousands of acupoints along the meridians and they are associated with specific organs or body systems. Nausea, for example, is treated by needles inserted into the wrists while visual problems might have needles inserted into the feet. Acupuncturists believe that the treatment stimulates the natural healing response. The insertion of the needles to a depth of between 0.5–2.5 cm is virtually painless. During the treatment there is often a tingling or heavy feeling around the needles which is an indication that the chi is being affected, but it is not painful. The needles used must be completely sterile and they are usually disposable, designed to be used only once. After the treatment many people report feeling greatly relaxed and invigorated (Figure 5.9).

Western science and acupuncture

Western science has never proved or accepted the existence of chi, but there is an increasing body of evidence which shows acupuncture leading to real changes in the physiology of the body (Figure 5.10). Today Western scientists believe that acupuncture may work by releasing chemicals such as endorphins and monoamines which block pain and change the blood flow in the brain and the rest of the body.

Acupuncture is widely used not only by people who have a clinical illness but also by people who simply feel 'below par' or generally out of sorts and unwell for no particular reason. The balancing of the chi is thought to restore energy and vitality and help to maintain 'whole person' health.

Another interesting piece of work involved a survey of 575 patients from six American acupuncture clinics during the early 1990s. These patients were suffering from a wide variety of complaints. The levels of satisfaction expressed were very high (Figure 5.11).

Scientists need evidence which will stand critical scrutiny by others in the same field, and it is also important for scientists that a plausible mechanism is proposed for any changes that are observed. This is another major stumbling block for complementary therapies, as many of the explanations involving 'energy fields', 'energy lines' or 'chi' do not fit in with the conventional model of how the human body works and heals itself. Complementary therapists would argue that these descriptions are simply another model for describing people, just as orthodox doctors would describe as Type A personalities people who have lots of aggression and drive and who are very stressed in their work.

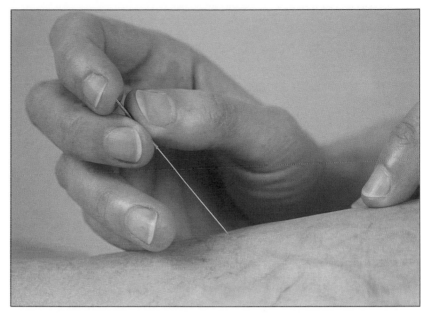

Figure 5.9

An acupuncturist about to insert a needle into the foot of a patient.

Question

11 How does the traditional explanation of how acupuncture works differ from the Western view of how the body fights disease and heals itself?

Ref to page 59

Figure 5.10 *A summary of some of the research findings on the effectiveness of acupuncture.*

Country of origin and year of the research	Summary of the main findings
FINLAND 1987	5 year trial of acupuncture therapy surveyed 348 patients treated with acupuncture for chronic pain. 65% of the patients with pain in the head, neck, shoulder or arm who had taken pain killers before treatment either stopped them completely or greatly reduced the dosage after treatment. Patients with osteoarthritis showed less effect.
CHINA 1980	Acupuncture treatment was beneficial to the majority of people with low back pain. 51% pain reduction was recorded in one treated group and 62% in another. The untreated group recorded no reduction in pain. After 40 weeks 58% of the treated individuals still felt the benefits of the treatment.
USA 1991	Acupuncture given to 21 patients with severe but stable angina. Their normal medication was maintained. During the period of acupuncture treatment angina attacks in the group dropped from an average of 10.6 to 6.1 per week. They could carry out more work before an attack was triggered and a lifestyle questionnaire showed that the patients had an improved feeling of well-being.
USA 1992	The effect of acupuncture on high blood pressure was measured on 30 patients. The immediate effect of the acupuncture was to lower the blood pressure on 100% of the patients, and in the long term 63% of the patients showed continued reduced blood pressure.

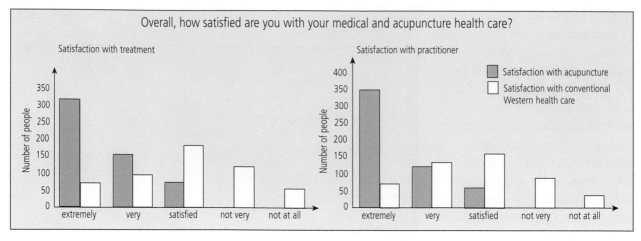

Figure 5.11

A comparison between the satisfaction of 575 Americans with their acupuncture treatment and their conventional medical treatment.

Questions

12 What conclusions do you draw from the information included in Figure 5.10?

13 How do you account for the low level of funding for research into complementary medicine?

14 Comment on the design of the study reported in Figure 5.12. What are its strengths and weaknesses? How convincing do you find the results? What conclusions do you draw from the results?

15 Doctors have understood for a long time that as many as a third of all patients respond favourably even when treated with a placebo (a non treatment) as long as they believe that they are receiving the real treatment. Could this placebo effect account for the results in Figure 5.12?

16 What features would you expect in a rigorous study of acupuncture?

One of the biggest problems for obtaining evidence about the effectiveness of complementary therapies is the low level of funding allocated to such research. For example, in the UK the NHS spends approximately 85p on investigating complementary medicine for every £100 it spends on conventional medical research. Pharmaceutical companies, the other big providers of medical research, are not interested in investing in complementary therapies because they usually involve techniques or herbs which are not highly profitable.

In spite of these difficulties a number of large research projects subjecting the use of acupuncture to investigations which will stand up to rigorous scientific scrutiny are currently underway. The two pieces of research described below, one carried out several years ago and the other now in progress illustrate the importance of such work.

Acupuncture and back pain

Back pain is the single largest cause of missed time at work in the UK. Ways of treating it include painkillers, muscle-relaxing drugs and physiotherapy. Many people also turn to acupuncture, and for some time there have been attempts to show that acupuncture really is a relatively cheap and effective way of helping back pain sufferers. In 1980 a research project was set up to investigate the effects of acupuncture on chronic lower back pain. The study was carried out on 56 men at a worker's clinic for lower back pain. These were people for whom standard treatments including in some cases surgery had failed. The group was divided into two. One group continued to receive standard therapy while the other was given standard treatment plus acupuncture at a variety of sites. They were given different numbers of treatments to a maximum of 15, and some were given electrical stimulation of the needles. The patients were fully assessed before the trial began, 12 weeks after the start of the treatment and just over 27 weeks after the treatment began (Figure 5.12).

This study appeared to show that a relatively simple course of acupuncture added to standard treatment gave a great improvement in the outcome of the treatment, with many more people able to return to work.

Figure 5.12

EFFECT OF TREATMENT	Patients who received standard treatment only (27)	Patients who received standard treatment and acupuncture (29)
Recovered to return to their original or equivalent job	4	18
Recovered enough to return to lighter work	14	10
Still unable to work	9	1

In the late 1990s a major research initiative was set up to look at the effectiveness of acupuncture as an option for the treatment of back pain. The York Back Pain Trial was set up by practitioners from the Foundation for Traditional Chinese Medicine working with GPs in York and the Medical Care Research Unit at the University of Sheffield. An initial feasibility study was done looking at the effect of acupuncture on four patients. A pilot study on 20 patients was then undertaken, and using the experience gained from this the methodology of the planned study was improved. An application for research funding from the NHS was put together and in 1998 the go-ahead was given for a trial based on 240 patients which would cost a little under £200 000.

The research involves a pragmatic randomised controlled trial. A pragmatic trial seeks to evaluate acupuncture as it is actually used rather than in an isolated laboratory environment. The trial is designed to answer the question 'Is it in the interest of patients for GPs to have access to acupuncture?' As part of this question the researchers recognise that acupuncture needs to be shown not only to be effective in the treatment of low back pain but also to be cost-effective. The numbers involved in the trial should give both clinically and statistically valid results.

The trial involves the randomisation of patients into two groups, both of whom will receive standard treatment for their back pain. However, one group of 160 patients will also be offered acupuncture. If they chose to accept they will attend an acupuncturist in the normal way and will be given treatment, advice on lifestyle and diet and build up a relationship with their acupuncturist as if they were normal patients. The research team are also looking at the economic implications of treatment – they are well aware that however effective acupuncture is, if it is much more expensive than orthodox treatments GPs will not be able to afford it for their patients. It is hoped that the trial will demonstrate whether or not acupuncture is of use to GPs in the treatment of lower back pain, and how effective it is at relieving the symptoms of those affected. The first patients were enrolled in the middle of 1999 and the results of the trial, and the conclusions drawn from the evidence, are expected in 2001–2. They will be published in a journal and also posted on the Internet where anyone who is interested will be able to see the results of this extensive investigation. In a paper about the York Back Pain trial, Hugh MacPherson, the Research Director of the Foundation for Traditional Chinese Medicine, says:

Questions

17

a How does the methodology of the York Back Pain trial compare with that of the trial described above (see Figure 5.12)? What are its strengths and weaknesses?

b How does the design of this study compare with your list of features from question 16?

18 In the York Back Pain Trial what were the purposes of the feasibility and pilot studies carried out before the full research proposal was drawn up?

19 How does a pragmatic, randomised trial compare with a double-blind trial? Why is a double-blind trial hard to arrange in acupuncture? What do you think are the advantages and disadvantages of a pragmatic trial?

Question

20

a Why is it so important that the cost of acupuncture as well as its effectiveness compared to orthodox medicine should be taken into account?

b If acupuncture were shown to be incredibly effective at treating lower back pain but also very much more expensive than normal treatments, what ethical issues would be raised by this? How would these issues be viewed by a GP with a practice to run and a patient with severe lower back pain which has lasted for 5 years and is not responding to conventional treatment?

'there are three outcomes which would be positive for acupuncturists; firstly that acupuncture is more effective and cheaper (than conventional treatments); secondly that acupuncture is more effective but not cheaper; and thirdly that acupuncture is as effective and cheaper. Whatever the outcome, costing all the economic dimensions will help to answer the question as to whether acupuncture should be more widely available as an option in primary care'.

What about the future?

The huge movement of ordinary people towards different forms of complementary therapy has forced the orthodox medical profession to look again at the choices on offer. Whilst some forms of complementary medicine are still regarded with suspicion or the assumption that any benefit is 'all in the mind', others are increasingly becoming accepted treatments. A new style of medicine called integrated medicine is slowly developing which takes the best features of both orthodox and complementary medicine. The complementary approach is not suitable for everything. If you are in a car crash, you need surgery, anaesthetics and drugs – a gentle aromatherapy massage will not help you much. Serious infections like meningitis need antibiotics fast, and failing hearts need bypass surgery. But for anyone suffering from chronic pain, eczema, allergies, severe morning sickness or back problems, then there may be a complementary therapy which will have some positive effects and may be less hazardous than any orthodox treatment while the body heals itself.

Review Questions

21 What does the study of complementary medicine suggest about the influence of a person's views and expectations on the data they collect in an investigation and their interpretations of it?

22 How do the interests and concerns of society influence the direction of scientific research and development in medicine?

Chapter 6

Genetic diseases

The issues

For many hundreds of years people have known that some diseases run in families. These genetic diseases are not infectious, but they can be passed on from one generation to the next.

Today it is possible to test people to see if they are carriers of a genetic disease. It is also possible to test to see if a fetus in the womb is likely to be affected by a disease inherited from its parents. The results of these tests can face people with agonising decisions.

At the moment there are no cures for genetic diseases, only treatments which can alleviate the symptoms. The new techniques of genetic engineering offer some hope that cures may be found but the prospect of gene therapy raises new dilemmas.

The science behind the issues

Each egg and each sperm cell contains a randomly selected half of each parent's genes. The single cell that they form at conception contains a full set of genetic information. The 'instructions' for development are made up of many pairs of genes with one of each gene pair coming from the mother and the other from the father.

Some inherited diseases are the result of a faulty form of a single gene. In these cases it is relatively easy to predict the chance of a baby suffering from the disease.

What this tells us about science and society

Like all scientific measurements, genetic tests are not perfect. There is always a chance that the results may be wrong. It is important to be able to estimate the reliability of a test.

New techniques of genetic engineering may make possible treatments which people find unacceptable. Just because something can be done does not mean that it ought to be done. Science cannot provide all the answers to these issues. They involve ethical decisions

There can be tensions between the rights of individuals and the interests of society. Some groups have campaigned to reduce the numbers of babies born with a genetic disease but in doing so they have had to restrict the freedom of individuals.

Figure 6.1

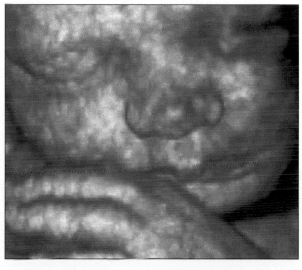

A coloured, three-dimensional ultrasound scan of the face and hands of a fetus. Genetic diseases are present from conception – as methods of investigating the secret world of the unborn fetus get better and better, it is becoming possible to identify problems before birth, but the options for action are very limited.

What are genetic diseases?

The primary source of Jewish law, the Talmud, specifies that if a baby boy dies of bleeding after circumcision, his younger brothers should not undergo the ritual and his male cousins on his mother's side are also exempt. This shows that, many centuries ago, there was a recognition that blood which does not clot runs in families and is inherited from the mother rather than the father. This strange condition is now known as haemophilia and it is an example of a genetic disease. Genetic diseases are not infectious. They cannot be 'caught' from another, infected person, but they can be passed on through the generations of a family.

Some genetic diseases are the result of problems with whole chromosomes. One of the most common problems of this type is Down's syndrome. People born with Down's syndrome have an extra copy of one chromosome (chromosome 21). This causes slow and limited

The gene model of inheritance

Figure 6.2

When the special sex cells are formed, the number of chromosomes is halved so each human egg or sperm contains only 23 chromosomes. The genes carried on the chromosomes are mixed and exchanged so each new sex cell has a slightly different mixture of genes.

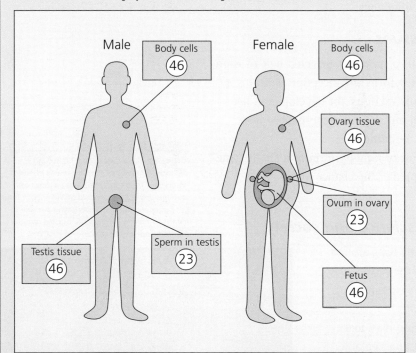

Figure 6.3

The Y chromosome is much smaller than the X and carries far fewer genes – in fact the human Y chromosome seems to carry little but male sex information and a gene which in one form can result in very hairy ears. This means that males are more likely to be affected by genetic diseases resulting from mutations on the X chromosome. As they only have one X chromosome, any problems in the genes will show in the appearance of the male offspring.

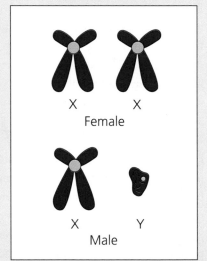

In the nucleus of every cell is the genetic material – pairs of chromosomes which carry the 'instructions' for the development of a new individual. This genetic information is carried in the form of many pairs of genes.

Normal human body cells (somatic cells) contain 46 chromosomes in 23 pairs. In sexual reproduction females make specialised sex cells called eggs and males make specialised sex cells called sperm. Eggs and sperm are germ-line cells (Figure 6.2).

In 22 of the 23 pairs of human chromosomes the two members of the pair are the same size and shape and carry genes for the same features. The one exception is the sex chromosomes

(Figure 6.3). Females have two X chromosomes, males have an X and a Y. Genetic diseases associated with the X chromosome are known as sex-linked diseases. Haemophilia is an example of sex-linked genetic diseases. Red-green colour blindness is also an inherited condition which is sex linked.

mental development and many physical problems including a lack of muscle tone which makes the face very inexpressive. Heart problems may also be a serious problem. Chromosome abnormalities are not usually passed from one generation to the next as the people affected are often sterile and often do not live long enough to reproduce.

Cystic fibrosis

Some genetic diseases are literally the result of one faulty gene. One person in every 25 in the UK carries the gene for cystic fibrosis, and about one in every 2500 babies born to white Europeans has the disease. It causes severe problems in the breathing and digestive systems linked to thick sticky mucus which the body cannot shift. It shortens the life of sufferers, who need physiotherapy several times a day and many drugs to keep their bodies functioning.

Both parents must be carriers of the disease for a child to have a chance of suffering from cystic fibrosis (Figure 6.4). Carriers are perfectly healthy because the normal gene is dominant. If a child inherits a faulty gene from both parents then the symptoms caused by the presence of the recessive gene appear.

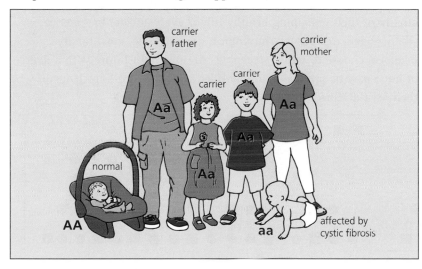

carrier father

carrier mother

carrier

carrier

Aa

Aa

Aa

normal

AA

Aa

aa

affected by cystic fibrosis

Huntington's disease

Huntington's disease is a genetic disease caused by a single dominant gene. This means that anyone who inherits the faulty form of the gene will get the disease, and that any children they may have will have a 50% chance of inheriting the disease too. Until very recently, the victims of Huntington's disease only knew they were affected when symptoms of the disease began to show. This is usually at between 30 and 50 years of age so many people have already had children before they discover their own disease. It is very rare, affecting about 1 in 20 000 of the white American population.

Huntington's disease involves the gradual destruction of nerve cells in the central nervous system. The person begins to have involuntary jerking or writhing movement of their arms and legs, and strange facial grimaces. Changes in personality occur, including laughing and crying at

Question

1 Why is it important that the number of chromosomes in the germ cells is only half of that in the normal somatic cells of the human body?

Key terms

Every individual carries two copies of each **gene**, one inherited from the mother and the other from the father

In the case of cystic fibrosis, the normal form of the gene is **dominant**. As long as a person inherits at least one normal form he or she will be healthy. A person has to inherit the faulty form of the gene from both parents for the symptoms to show up. The faulty form of the cystic fibrosis gene is **recessive**.

Figure 6.4

Inheritance of cystic fibrosis. Both parents are healthy. They each have one normal (A) and one faulty gene (a). Each child has a one in four chance of inheriting two faulty genes and showing the symptoms of the disease.

Questions

2 Is there any chance of a child inheriting cystic fibrosis if only one parent is a carrier?

3 How does an inherited disorder such as cystic fibrosis differ from an infectious disease such as measles?

4 How many of the 50 million white people in the UK are carriers of cystic fibrosis if 1 in 2500 children inherit the disease?

the wrong time, inappropriate anger, memory loss and bizarre behaviour. The pattern of symptoms varies a lot, but the patient usually loses the ability to communicate several years before their inevitable death.

In 1983 Dr James Gusella working in Boston, Massachusetts decided to compare the DNA of family members affected by Huntington's diseases with those who did not have the disease. He discovered a 'genetic marker' present only when the Huntington's gene was there. This discovery was the basis for a genetic test which has been developed for use with family members when one member of the family develops Huntington's disease. In the past people faced many years of uncertainty before knowing if they had inherited the disease. The test opens the door for people to face and plan their own future, and also to plan whether to have children. For example, if one partner has the Huntington's gene, a couple may chose not to have children and thus risk passing on the gene. On the other hand they may have children, but screen the fetuses, ending the pregnancy if the Huntington's gene is present.

At first sight such knowledge appears invaluable, but some people might find it unbearable to learn that they have inherited a completely untreatable fatal disease, leading them to depression, a greatly reduced quality of their remaining healthy life or even suicide. In America, where genetic testing for Huntington's is available, most family members choose to refuse screening. An uncertain future with a degree of hope may be preferable to one filled with the fore-knowledge of their own decline and death.

Questions

5 What is the chance that a baby will be affected by Huntingdon's disease if one parent carries the faulty form of the gene but the other parent has two normal forms of the gene?

6 Many genetic diseases show themselves within the early years of a child's life. How does Huntington's disease differ? Do you think this makes it better or worse for the families concerned?

Figure 6.5

Because the symptoms of Huntington's disease do not appear until early middle age, and because the genetic nature of the disease was only recognised relatively recently, the disease has cut a path through families for generations, as can be seen clearly in this American family tree.

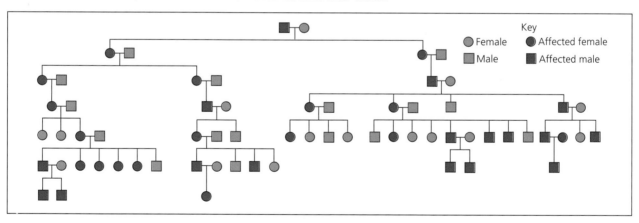

Question

7 Suppose that you believe that the human fetus has a right to live. What duties does this mean that a pregnant women has to her fetus? Does anyone else, such as the father, other members of the family or society in general, have duties to her fetus?

'Is the baby all right?'

For most of human history the only way to find out if a baby is 'all right' has been to wait and see, and for much of the world this is still the case. But the technology now exists to find out an amazing amount about a baby before it is born. Antenatal testing is becoming more and more important in the care of pregnant women. It is used not only to pick up genetic problems in the developing fetus but also to identify problems in development, where the fetus fails to form properly for a variety of reasons. An example of such a problem is spina bifida, when the backbone fails to develop properly, perhaps as a result of a lack of folic acid in the mother's diet.

The tests

Blood tests relatively early in pregnancy pick up women who are at risk of having a baby with Down's syndrome. Tests are also available for other genetic diseases. The simplest methods involve blood tests and ultrasound scanning. Ultrasound is non-invasive and shows if the fetus is developing normally. But if there is real concern it is necessary to obtain a sample of fetal cells to make up a complete picture of the chromosomes and see if there are any problems. To obtain the material one of two tests is carried out – amniocentesis or chorionic villus sampling (Figure 6.6).

Question

8 What are the advantages and disadvantages of amniocentesis and chorionic villus testing?

Tests for genetic diseases

Figure 6.6

Techniques like these make an accurate diagnosis of Down's syndrome and other genetic disorders possible before birth.

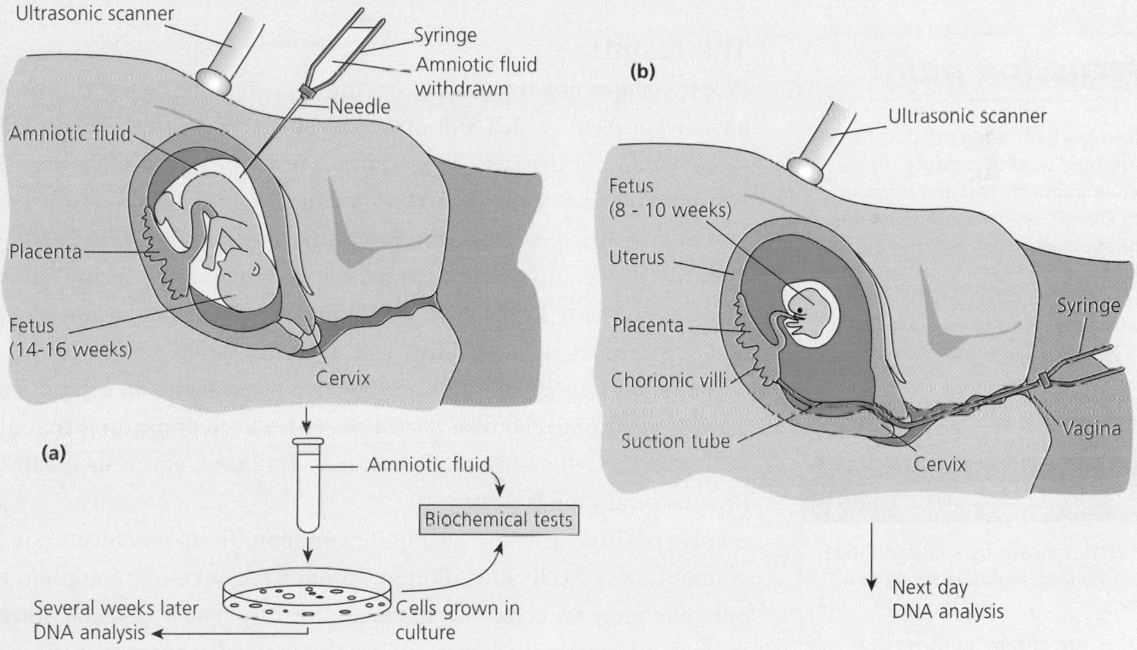

AMNIOCENTESIS

This involves removing about 20 cm³ of the amniotic fluid which surrounds the fetus. It is withdrawn using a needle and syringe and is carried out at about the 14th to the 16th week of pregnancy. Fetal cells can be recovered from the fluid and after they have been cultured for several weeks a number of genetic defects as well as the sex of the baby can be determined from examination of the chromosomes from these cells.

DISADVANTAGES:
- The test can only be carried out relatively late in the pregnancy.
- The results are only available several weeks after the test.
- The test carries about a 0.5% risk of miscarriage after the procedure, whether or not there is something abnormal in the genetics of the fetus.

CHORIONIC VILLUS SAMPLING

This involves taking a small sample of tissue from the developing placenta. This makes a much bigger sample of tissue available for examination. The fetal cells are then tested for a wide range of genetic abnormalities. This diagnostic technique can be carried out earlier in the pregnancy (at 8–10 weeks) and it gives the information more rapidly.

DISADVANTAGES:
- There is a 1.5% risk of a miscarriage after the procedure whether or not there is something abnormal in the genetics of the fetus.

All X chromosomes from the father are inactivated in fetal placental cells so any problems in the genes on that chromosome cannot be detected by this technique.

Some people refuse to have antenatal testing carried out. Why do you think this might be?

Suppose that you take the view that to decide on the rightness or wrongness of a course of action it is enough to look at the likely consequences of the action. Suppose too that you and your partner are trying to decide whether to use amniocentesis to see whether the 15- week-old fetus one of you is carrying has Down's syndrome.

 Ref to pages 35–37

Make lists of the following and discuss what you should do:

- the positive consequences that might follow if you use amnio-centesis;

- the negative consequences that might follow if you use amnio-centesis;

- the positive consequences that might follow if you do not use amniocentesis;

- the negative consequences that might follow if you do not use amniocentesis.

Discussion point

Recent evidence suggests that the blood tests used to indicate that a woman might be carrying a fetus with Down's syndrome are no more accurate than simply looking at the age of the mother and carrying out a normal scan. What will be the advantages if this turns out to be true? What, if any, disadvantages can you foresee?

Questions

9 What is meant by a false positive and a false negative result to a test?

10 The rate of false positives or false negatives matters as does the level of incidence of the genetic disease.

a If a condition affects 1 in a 1000 babies and the test has a false positive rate of 10% and a false negative rate of 5% is it worth using?

b Is it more important to develop the technique to reduce the rate of false positives or the rate of false negatives?

c For a condition with a frequency of 1 in a 1000 cases, how low should the rates be before it is clearly justifiable to use the test?

The results

People tend to assume that by having a medical test they will be given a clear-cut result which will make the situation clearer. Unfortunately this is often not the case. Sometimes a test will give a **false negative**. For example, many pregnancy tests can be carried out as soon as the monthly period is missed. However, a negative result at this early stage does not necessarily mean a woman is not pregnant – it may simply be that her hormone levels have not built up sufficiently to be picked up and she needs a more sensitive test, or to try again a week later.

The blood tests which indicate the risk of problems in a fetus such as Down's syndrome can also give a false negative, indicating that all is well with the fetus until later tests may find problems – or a baby with Down's syndrome is born.

False positive tests are also quite common. In an amniocentesis test several tubes of cells are cultured. In one case, a couple were told that one tube grew no cells, one showed a healthy female and the other tube showed a female with a rare and fatal genetic disorder – yet there was only one fetus. The pregnancy was monitored throughout, with plans to end it if the baby showed deformities on the scans. Finally a completely healthy baby girl arrived, after months of anxiety and much expensive and unnecessary care. This situation is not uncommon. Even tests for diseases such as Huntington's disease are not foolproof – people can be tested and given false results, or an uncertain result, so that they still do not know their future risks.

So tests are rarely clear cut. They have varying degrees of uncertainty. It is important to recognise this and take account of it, when such results are being used to decide whether a pregnancy should be terminated or continued to allow a baby to be born.

What about abortion?

Many pregnancies where the fetus has a genetic disease end in abortion. Natural abortion when the fetus dies and is lost from the body (also called a miscarriage) is very common in the first 3 months of pregnancy and in many of these cases the foetus has some sort of genetic problem.

However, sometimes a pregnancy is deliberately ended in a medical abortion. Legally these are carried out because there is a risk to the health of the mother or if the baby is handicapped in some way. In many cases, however, the real reason for the abortion is that the pregnancy has not been planned and the baby is not wanted. Most abortions are carried out in the first 3 months of pregnancy, when the embryo is still very small and a long way from being capable of independent life. However, in many countries abortion is legal up until 20 weeks or even later. This is a traumatic process as birth is induced and the woman has to deliver the fetus, and it is usually only used when serious genetic handicaps are picked up by amniocentesis testing late in the pregnancy.

The ethics of abortion causes great divides. On one side are people who believe that women have a right to decide if and when they wish to have a baby, and that every child born has the right to be a healthy and wanted child. While everyone would prefer it if abortions were not necessary, these people feel that if a woman is pregnant and does not wish to be, or if the fetus has a genetic disorder or some other developmental problem, then a termination of that pregnancy is the right course to take.

Other people feel equally strongly that it is never right to take a human life, even that of a few-weeks-old embryo. They believe that the rights of the unborn should be equal to those of the parents and any other siblings who might already be in the family, and that abortion is always wrong.

These are issues of great sensitivity which people feel very strongly about.

Difficult choices

When antenatal testing shows that a fetus has a genetic disease it can be very hard for parents to come to terms with the fact that the child they are expecting will not be the 'normal' healthy baby they had hoped for. The choices facing them are often stark. At the moment there are no cures for many of the genetic diseases which can be identified by testing. The only choice is to end the pregnancy by an abortion or continue with it and, if the child survives birth, look after it and cope with the genetic disease.

The parents may choose to terminate the pregnancy rather than condemn the child to a life of handicap or illness and perhaps prolonged and painful death. The parents have to make judgements about the quality of life of an individual who is not yet born. This is very difficult to do. Some people think even a mild handicap would be devastating, while other people live enormously full and rich lives in spite of terrible disabilities. Parents also have to weigh in the balance the effect a genetically damaged child would have on their own lives and on the lives of

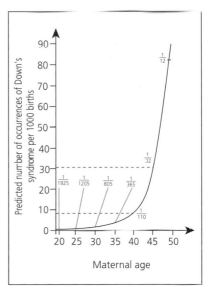

Figure 6.7

The likelihood of having a baby with Down's syndrome increases with the age of the mother. As many women in developed countries are delaying having children until they are well established in a career, more and more will inevitably face difficult choices when problems arise.

11 Why is it important to ask about the reliability of any antenatal test you might be offered?

12 A family is told they have a 1 in 4 chance of having a child with cystic fibrosis. Does this mean that:

- if they have four children one will definitely inherit the disease,

- if they have three normal children already, the fourth will definitely have cystic fibrosis,

- if their first child inherits the disease, then their next child will not,

- if their first child does not have the condition, the chance of the next one having it is greater than 1 in 4?

Discussion point

A young couple expecting their first baby have been told that the fetus has Down's syndrome. The child may be severely mentally and physically handicapped, and never be capable of looking after him- or herself, or it may have a relatively minor level of handicap and, with plenty of support throughout its life, play a useful role in society.

What are some of the arguments for and against abortion that the young couple might consider?

Key terms

Genetic testing involves the testing of the members of a family into which a child with a genetic disorder has been born. The aim of testing is to estimate the risk of further affected babies and to identify the people who are carrying the faulty gene.

Genetic screening involves the genetic testing of a large proportion of the reproductively active population for a particular faulty gene.

any other children they might have – many feel unable to shoulder such an unasked for burden yet others find their lives enriched by the experience. Added in to this are all the moral and ethical issues associated with abortion. People face a terrible dilemma where there is no easy solution.

Genetic counselling

For most people finding out about genetic diseases in their family is very traumatic. Overnight they change from seeing themselves as normal healthy people who will go on and produce a normal healthy baby to realising that they carry a gene which could make them (or more probably their children) seriously ill or handicapped. All of the issues discussed above are suddenly of immediate and personal relevance. Because for many couples the first realisation they have of a problem is when they are pregnant or shortly after their baby is born, the trauma is even harder to bear.

Genetic counsellors are trained to help people come to terms with the situation of carrying an abnormal gene which can cause genetic disease. A family pedigree will be worked out, which may be useful in confirming the diagnosis and can also be used to indicate any other individuals who might unknowingly be affected as carriers. This, however, raises another difficult issue. Should those other potential carriers, some of whom may be quite distant relations, be told or not? Genetic counsellors will help people come to a decision about issues like this. They will also assess the statistical risk of a couple producing another child with the same defect. If the problem is caused by a single gene then the risk can easily be calculated. Even if it involves lots of genes or a whole chromosome abnormality the risk can be calculated using accumulated medical data.

For most genetic defects, the only real options open to couples are to avoid having children (or any more children), to undergo antenatal screening and abort affected pregnancies or to have children without screening and hope for the best. Again the role of the genetic counsellor is to help couples recognise the options and work their way through to the alternative which is right for them within their own framework of moral, family, religious and social beliefs and traditions. There are as yet no easy solutions to this problem. All that can be hoped for is that a couple reach as comfortable accommodation with the facts as they can. Whether the advances promised by techniques such as gene manipulation will bring real help and relief to families affected by genetic diseases is still to be seen. The concern is that the issues such techniques raise might be just as difficult for individuals and society to resolve.

Are there any solutions?

Couples who find they are at risk of producing children with genetic defects have several options open to them. They can go ahead and have a family as usual, hoping that in the genetic lottery they are lucky and their children inherit healthy genes, but prepared to support and take care of them if they do not. They may decide not to have children at all, to prevent passing on a faulty gene even to a carrier. The third option is to go ahead with pregnancies but to have each pregnancy screened and terminated if the fetus is affected (Figure 6.8). This option can be very traumatic, and is not open to some people because of their beliefs about abortion but it is effective at preventing the birth of children with genetic diseases. What it doesn't do is weed out individuals who will be carriers, so the gene continues into the next generation.

A more sophisticated version of this technique involves fertilising the ova of a couple at high risk of producing offspring affected by genetic disease outside the body.

Only those embryos free of the problem genes are placed in the mother's uterus to implant and grow. For example, in the case of sex-linked diseases such as haemophilia only female embryos would be replaced.

Another aspect of this complex picture is that with techniques of genetic engineering, the possibility of inserting a healthy gene into the cells of an individual with a genetic disorder is becoming a possibility. Trials are already going ahead in people with cystic fibrosis, providing cells which can manage sodium and chloride ions appropriately and so allow normal, healthy mucus to be produced, and the results show some promise. ⊖ *Ref to Chapter 8*

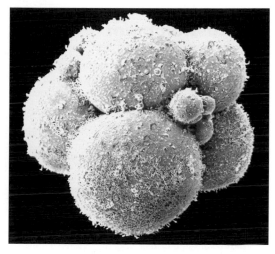

Figure 6.8

Coloured electron microscope picture of a human embryo at the eight cell stage about three days after fertilisation. This is magnified about 900 times. When a human embryo is a small ball of cells, a single cell can be removed to check the genetic makeup without causing harm to the development of the embryo.

Another disease which is being considered for this sort of treatment is haemophilia. A virus has been used to carry the healthy gene for the missing blood clotting factor into the tissues of haemophiliacs. The gene started working, and the human 'guinea pigs' needed less artificial clotting factor as their bodies began to make the missing protein. There is a long way to go but this could be the beginning of an effective cure. But as with most new technologies, there are more ethical dilemmas here.

At present, genetic manipulation of the germ cells is completely forbidden in the UK and most other countries. This means an affected individual may be helped in their lifetime to be free from the effects of the genetic disorder they have inherited but they will nevertheless still be at risk of passing it on to their offspring. Treatment has to be ongoing through the generations, which is an expensive option. It can only be a matter of time before it is suggested that manipulation of the germ cells as well as the somatic cells would treat not just the individual concerned but all their potential future descendants too. And while this sounds almost miraculous in theory, once it is permitted to manipulate the genes of the germ cells for medical reasons there are those who fear that pressure would increase for changes to be made in the genes for intelligence, or beauty, or height or aggression. The days of eugenics, of manipulation of the human breeding stock and the creation of sets of 'designer humans', might then become not science fiction, but science fact.

Questions

13 What is the difference between gene therapy of the somatic cells and gene therapy which involves the germ-line cells?

14 Is there an ethical difference between the two types of therapy, when both involve manipulation and changing the genetic material?

Genetic screening

Genetic testing and counselling is designed to help families. It is up to individuals to decide whether they want to be tested and then to seek advice about the results.

Genetic screening is part of a health policy to reduce disease. Governments, health authorities and whole communities decide whether or not to introduce large-scale screening programmes.

At first sight the idea of genetic screening seems an excellent idea – identify the people with faulty genes (and everyone has some) before they pass them on to any children they might have in the future. But what are the implications of this type of knowledge? What do individuals, doctors and the authorities actually do with it? And can genetic screening as it stands give us all the information we need? These are questions which need to be addressed, because there is currently a major research drive known as the Human Genome Project which aims to produce detailed maps of the whole of the human genetic material. As this project moves forward it will be possible to identify the genes responsible for all the single gene disorders, and to provide screening tests for them. It should also make clearer the situation when genes and lifestyle both affect the health – for example, in conditions such as heart disease.

As discussed above in the context of Huntington's disease, knowledge of one's own genetic problems can be hard to bear. Another problem is who should have access to all of this information.

If an individual is screened and found to have a gene for a specific disease or to be a carrier of a problem gene, then one possibility is for the information to be confidential to them. Each person could then have to decide whether to tell partners, parents and friends, whether to have children and whether to tell their insurance company.

On the other hand some people argue that society has a right to know. If the knowledge is there, people should know that their partner will not knowingly pass on abnormal genes, and financial institutions should not be tricked into lending money or offering cheap life insurance to someone who knows they carry a genetic timebomb.

However, on these arguments every one of us is a bad risk. Most individuals will not carry the gene for Huntington's disease – but in the UK 1 in 25 carry the gene for cystic fibrosis. Should this affect our

---

Discussion point

Would you want to know your own genetic makeup? How would you use the information if you had it?

What guidelines would you like to see in place before everyone has access to his or her own genetic information?

The fall of the genetic dice

The choices for parents are relatively clear-cut when the inheritance of a particular gene or pair of genes results in a specific genetic disease which is shown up by screening. But the situation is not always so clear. Many 'lifestyle' diseases such as heart disease and many cancers are **multifactorial** – they are triggered when a number of different factors are in place. One common factor for these diseases is a genetic component. For example, some families have a genetic tendency to have problems in the way they metabolise fats, leading to furring up of the arteries and possible heart attacks. However, this genetic tendency may never show itself if the members of the family all eat a very low fat diet and take lots of exercise throughout their lives.   *Ref to Chapter 4*

Other families have a genetic predisposition towards developing cancer. We all hear stories of people who have smoked 60 cigarettes a day and lived until they are 90 – which tells us they do not have a genetic tendency to cancer – while other people who have smoked relatively lightly die in their forties from lung cancer, because their genetic make-up means that the growth control systems in their cells are easily tipped out of balance by the cancer-causing chemicals in cigarette smoke.

The issues for these inherited tendencies are rather different from those raised by straightforward genetic diseases because they may never reveal themselves if other factors are not in place. The difficulty is that once knowledge is available it can be very hard to ignore.

Questions

15 What are the advantages of genetic testing and screening?

16 What are the disadvantages of genetic testing and screening?

choice of partner? And many more people will carry genes which mean that, if certain environmental conditions are met – for example, they smoke, or eat too much – they will be far more likely to develop heart disease or cancer than someone else with exactly the same habits but different genes. It is possible that an individual with genes loading for heart disease might be refused a driving licence after the age of 25 because of the increased likelihood of their dying at the wheel and causing a serious accident. If it is known that a young person is at high risk of serious disease later in their life, will they be disadvantaged when they look for college places, loans, jobs and relationships? These sorts of scenarios can sound a little far-fetched, but it is better for society to consider the possibilities now and set some guidelines in place than wait until the situation is happening.

How far should we go?

It can certainly be argued that it is unfair to have children affected with a serious genetic disease when the knowledge is there to prevent it. But knowledge does not solve the problems, it simply raises further dilemmas. Should all couples who plan to start a family or indeed, who plan to have sex (as there is always a risk of pregnancy unless one of the partners is sterilised) be screened to see if their genetic weaknesses are compatible? Among the Ashkenazi Jews of North America, a genetic disease called Tay-Sachs disease is very common.

In an effort to reduce the incidence of Tay-Sachs disease, the rabbis have organised screening of each couple who plan to marry. If it is found that they are both carriers of the Tay-Sachs gene then the marriage is forbidden. The human cost for those couples who find they are genetically incompatible must be great, yet the saving of human misery if Tay-Sachs can be reduced in or removed from the population would be enormous. In the North American community the incidence of the disease has already been lowered by about 80%.

Screening all couples at risk of carrying a particular gene can however be enormously expensive. In a case like cystic fibrosis screening would need to be done over the whole reproductive population of Europe and then only 70-75% of the possible cases of cystic fibrosis would be prevented. The economic value of screening programmes can be looked at to see if they are worth carrying out in economic terms. For example, in 1979 it was estimated that to find one baby suffering from a genetic disease known as phenylketonuria (PKU), and then supply it with a special diet until adulthood, 15 000 babies had to be screened at a cost of £20 000. However, the lifetime treatment of one untreated, severely handicapped PKU baby cost £126 000 so the cost of the screening of all new-born infants for PKU was well justified from an economic point of view.

For cystic fibrosis, the benefit of whole population screening would be much smaller in relation to the cost than in the case of PKU. The cost of screening 2000 fetuses to find and abort the one with cystic fibrosis is

Discussion point

Explain what is meant by autonomy Does the Tay-Sachs screening programme enhance or diminish the autonomy of a couple who are planning to marry? Discuss the arguments for and against this particular screening programme.

Key term

Phenylketonuria (PKU) is an inherited defect in which the amino acid phenylalanine cannot be metabolised by the body. Levels of phenylalanine build up in the blood and damage the brain, causing very severe handicap. If phenylalanine is avoided in the diet during babyhood and childhood, brain damage can be completely avoided and during adulthood the brain loses its sensitivity to the amino acid so a normal diet can be eaten.

Questions

17 Why are all new born babies in the UK screened for the genetic disease PKU when there is no national screening programme for cystic fibrosis?

18 Why is there no national screening programme for Huntington's disease?

19 Do you think that it is ethical to have screening tests available for a number of diseases but not to use them because of the expense?

little less than the cost of treating an individual suffering from cystic fibrosis throughout their shortened life (Figure 6.9). As a result, cystic fibrosis screening is kept to families where the disease has occurred.

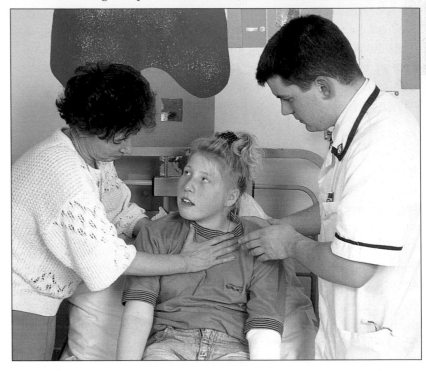

Figure 6.9

A physiotherapist teaching a mother of a young person with cystic fibrosis how to carry out a massage to relieve the build-up of mucus in the girl's lungs. Economic calculations look only at the financial costs of care. The human suffering of affected individuals and their families is not, and probably cannot be, quantified.

Review Questions

20 In the future it is quite possible that everyone will have access to their own genotype (the genes they carry). This raises a number of questions about who else should have access to this information. Consider each of the following:

a You and your partner are planning to start a family. You know that you have a gene which could cause a genetic disease in your children if your partner has a similar gene. So far you have not shown each other your genetic information. What do you think you should do? Is it the same as you suspect you actually would do?

b You are due to set out on a very long haul plane flight. Would you want to see the genotype of the pilot and crew of the aircraft? What information might it give you and would this be helpful?

c A research post is advertised for someone to be part of a team carrying out long-term experiments into human ageing. One applicant (who is now aged 25) stands out as having the ideal qualifications for the job and being the right sort of personality to fit in with the rest of the team. However, his genetic make-up shows that he has an increased risk of suffering from mental health problems in his forties. Should this affect his chances of getting the job?

21 Justice is about fair treatment and the fair distribution of resources or opportunities. If screening for genetic diseases becomes widespread, do you think people should have to pay for the tests or should they be provided free by the National Health Service? Justify your views in terms of the concept of justice. *Ref to page 35–37*

Chapter 7

Genetic Engineering

The issues

The introduction of genetically modified (GM) crops has become highly controversial in the UK and some other parts of the World. The principal objections concern possible harm to human health, damage to the environment and unease about the 'unnatural' status of the technology. Many people in the UK object strongly to the imposition of a new and untested technology which does not appear to offer them obvious benefits. Meanwhile there are those who strongly believe that genetic modification will prove vital for securing food supplies in parts of the developing world.

The science behind the issues

Genetic engineering rests on the principle that all organisms, however different they may appear to be, have in common certain fundamental similarities at the level of their genetic material. Because of this, genes can be moved by scientists between completely unrelated species, for example between humans and yeasts or between bacteria and plants. Furthermore, the gene operates in its new host in much the same way that it did in its old host.

Figure 7.1

Protesters demonstrating against genetically modified organisms.

What this tells us about science and society

Genetic engineering is not just a science; it is an industry. Industry both uses science and drives its development. Science, though, is often unpredictable, and technologies, such as genetic engineering, may have unintended consequences. Some see the motives of industry as unduly dominated by profit. People's views are coloured by the extent to which they trust the statements of scientists, industrialists and politicians.

Some people acknowledge potential gains of genetic engineering, but are aware of possibly catastrophic side-effects. Adopting the 'precautionary principle', they would argue that we should err on the side of caution until risks can be more reliably assessed. Work in this area needs to be regulated. However, it can be difficult to get agreement on how this should be done.

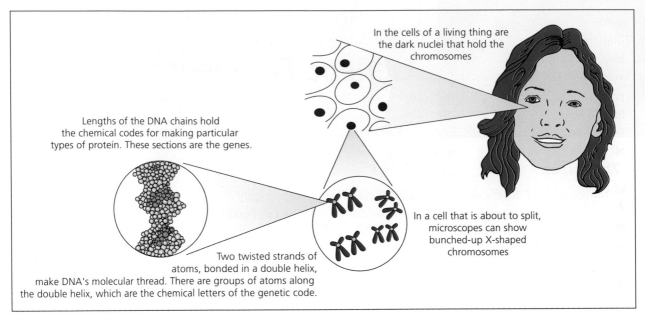

In the cells of a living thing are the dark nuclei that hold the chromosomes

Lengths of the DNA chains hold the chemical codes for making particular types of protein. These sections are the genes.

In a cell that is about to split, microscopes can show bunched-up X-shaped chromosomes

Two twisted strands of atoms, bonded in a double helix, make DNA's molecular thread. There are groups of atoms along the double helix, which are the chemical letters of the genetic code.

Figure 7.2

Genes are sections of the long DNA molecules in the chromosomes. The genetic code is a chemical code that can store and transmit information.

Figure 7.3

The basic mechanism of genetic engineering.

The basic principles of genetic engineering

The whole of genetic engineering relies on the fact that the genetic make-up of all organisms is fundamentally the same. Bacteria, fungi, plants and animals all have DNA as the material that makes up their genes (Figures 7.2 and 7.3).

Genes are written in a chemical code along the length of DNA molecules. There are four molecular 'letters' in the code and they are the same in all organisms. Because of this underlying similarity in the genetic code, it is now fairly easy for scientists to move genes from one

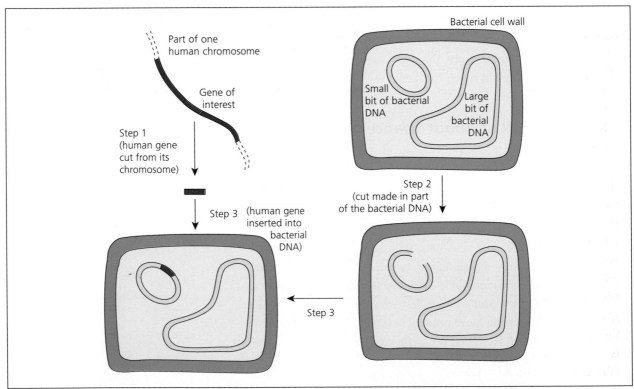

Part of one human chromosome

Gene of interest

Step 1 (human gene cut from its chromosome)

Step 3   (human gene inserted into bacterial DNA)

Bacterial cell wall

Small bit of bacterial DNA

Large bit of bacterial DNA

Step 2 (cut made in part of the bacterial DNA)

Step 3

species to another. Various methods are used but the basic principle is illustrated in Figure 7.3. Suppose you want to move a gene from a human to a species of bacterium. First, you need to identify the gene you want to move. Actually, this is the hardest part of the whole procedure. After all, species have tens of thousands of genes (humans have about 50 000), so finding the right one is a bit like trying to find a friend's home when you have forgotten his or her name and address and all you can remember is that the house is somewhere in Bradford.

Once you have identified the gene you want, you can use a special type of enzyme to cut the gene out from where it normally is (step 1 in Figure 7.3). The same enzyme can then be used to make a cut in part of the genetic material of the bacterium (step 2). Finally you use a different enzyme to insert the human gene into the bacterium's DNA (step 3).

Basically, that's it. The bacterium now has a human gene inserted into its DNA. DNA makes proteins, so when the bacterium now makes its own proteins it will also make the human protein coded for by the human gene. Suppose, for example, that the human gene was the gene that coded for human insulin, the hormone that helps to control the amount of sugar we have in our bloodstream. The bacteria would now make human insulin.

In fact, genetically engineered insulin is one of the few definite success stories of genetic engineering. Many people need injections of insulin two or more times a day because they don't produce enough of their own (Figure 7.4). These people have insulin-dependent diabetes. Before the advent of genetic engineering, the necessary insulin was extracted from the pancreases of slaughtered pigs or cows. Some people had ethical objections to this. In addition, there are minor but significant differences between the structure of human insulin and that of pigs and cows, which sometimes caused problems. Finally, there were occasions when viruses or other disease-causing organisms were inadvertently transferred to people from cows or pigs.

Genetically engineered human insulin was developed in 1980. Within a few years most of the insulin used by people with insulin-dependent diabetes was genetically engineered.

Nowadays there is a variety of ways of moving genes from species to species in addition to that shown in Figure 7.3. One surprisingly simple way is to fire the genes, using a tiny gun, at the organisms you want to genetically engineer. This technique is widely used for the genetic engineering of certain crops. Another way is to use a virus or a bacterium to carry the gene from one species to another. Bacteria are often used in the genetic engineering of crops while viruses are being tried in research intended to use genetic engineering to carry healthy genes into humans who have defective genes. A final way is to inject the DNA directly into the fertilised egg of an organism. This technique is widely used for the genetic engineering of mice, sheep and other laboratory or farm animals.

Figure 7.4

Genetically engineered human insulin has contributed significantly to the quality of life of people with diabetes.

Key terms

Genetic engineering usually means moving genes from one species to another. The term also covers the insertion into organisms of artificial genes – i.e. genes made in the laboratory. Other terms for genetic engineering include 'genetic modification', 'genetic manipulation' and 'recombinant DNA technology'. Genetically engineered organisms are sometimes referred to as 'transgenic organisms'.

Biotechnology is the application of biology for human purposes. It involves using organisms to provide us with food, clothes, medicines and other products. Traditional biotechnology includes the breeding of crops and farm animals and the use of yeasts in cheese-making, brewing and wine production. Modern biotechnology uses such techniques as genetic engineering, cloning and embryo transfer.

Agriculture means the growing of plant crops such as wheat, maize, rice, potatoes and melons.

Traditional agricultural breeding programmes

Agriculture has been around for some 8000 – 10 000 years. The birth of agriculture in various parts of the world was marked by the start of cultivation of wheat and barley in the Middle East, potatoes and beans in Peru and rice in Indochina. Agriculture requires the following four stages:

- sowing of seeds;

- caring for the plants;

- harvesting;

- selecting and keeping back some of the seeds for the next generation.

Traditional agriculture uses some sophisticated breeding methods. We now know that today's wheat, for example, is the result of crosses between different species of grass performed by people many thousands of years ago. One frequently used technique in traditional agricultural breeding programmes is to cross (breed) two different varieties (or sometimes two closely related species). Farmers then select among the offspring for the characteristics they want (Figure 7.5).

Suppose, for instance, farmers want a variety of rice with a high yield and also resistance to a particular species of parasitic fungus. And suppose that there are already some varieties of rice with a high yield and others with good resistance to the fungi. Breeders would probably cross one of the highest yielding varieties with one of the most resistant varieties and select among the offspring for individuals that had both features. Assuming that they manage this, they then need to spend several years 'bulking up' the plants to make sure there are enough of them not just for research but for full-scale farming. All in all, it

Figure 7.5

A researcher removing the reproductive part of a wild wheat flower to cross pollinate with a strain of wheat commonly used in agriculture. The aim is to produce new strains which are more disease resistant or able to grow well under different environmental conditions.

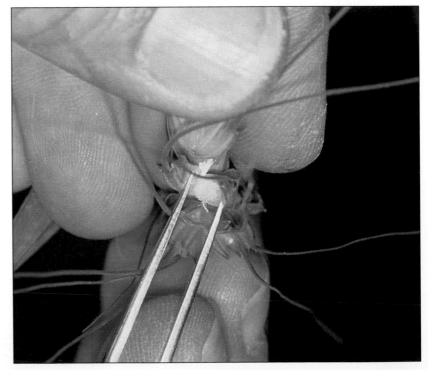

typically takes about 6 to 10 years to come up with a new crop variety using traditional breeding programmes. Genetic engineering provides a faster way of developing new crop varieties.

Applications of genetic manipulation in agriculture

Genetic manipulation is rapidly being introduced into certain crops (Figure 7.6). In 1999 around half of all the soya and maize grown in the USA was genetically engineered. There are two main reasons why crops are genetically engineered: to make them herbicide-tolerant or to make them resistant to pests. Different varieties of maize (known in the USA as corn) have been bred for both these reasons so we shall concentrate on maize here. However, the basic story is much the same for soya, sugar beet and an increasing number of other crops. Because of the commercial importance of agriculture, biotechnology both draws upon what is known about plants and their genetics and facilitates the emergence of new scientific knowledge thanks to the large amounts of money invested by commercial companies.

Pest-resistant maize

Some 7% of potential world maize production is lost to a moth called the European corn borer (Figure 7.7). This equates to 40 million tons of maize a year, worth approximately US$2 billion. Of the order of US$40 million worth of conventional pesticides are currently used each year to stop the pest doing even more damage. However, losses are difficult to eliminate by the use of such pesticides largely because the larvae live inside the maize stalks. Biological control (e.g. through the use of parasites or predators of the moth) has not been very successful.

Maize has recently been genetically engineered by a number of companies so that it produces the Bt protein of the soil bacterium *Bacillus thuringiensis*. When the European corn borer larvae eat this protein, their intestinal walls are damaged, causing them to die from hunger. Such transgenic maize has an average yield increase over conventional maize of around 5 –10% in the USA.

Question

3 Suppose one breed of cattle has a low incidence of lameness but a low milk yield whereas another breed of cattle has a high milk yield but a high incidence of lameness. Describe how traditional breeding might be undertaken to produce a new variety with a high milk yield and low incidence of lameness. (Don't forget that male cattle – bulls – don't produce milk.)

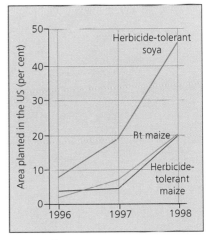

Figure 7.6

The late 1990s showed a dramatic increase in the planting of GM crops in the USA.

Question

4 With reference to crop breeding, summarise how commercial companies both draw upon scientific understanding and facilitate its development.

Figure 7.7

A caterpillar of the European corn borer (Ostrinia nubilalis) feeding on a maize stem.

Herbicide-tolerant maize

Weeds are plants that grow where we don't want them. There are many reasons why weeds are considered undesirable. The most frequent is that they compete with a food crop for nutrients, water or space. Weeds can be dealt with in a number of ways – where labour is cheap they can be removed by hand, for example. In many countries, though, herbicides are extensively used to control weeds. Indeed, approximately 90–95% of the area of land used to grow crops in Europe and the USA is treated with herbicides each year (Figure 7.8).

At present a large number of different herbicides are available commercially. Not surprisingly these differ considerably in their environmental impact - for example, their biodegradability and effects on non-target species such as insects, birds and mammals. Genetic engineering holds out the hope that instead of researchers starting with desirable crops and then finding herbicides that kill the weeds without harming the crops, they could start with the most desirable herbicides and then genetically alter the crops so that they, unlike their weeds, are unaffected by the chemicals.

Figure 7.8

Spraying herbicide amongst rows of grape vines. Over 90% of European and North American agricultural land is currently sprayed with herbicides so as to reduce the number of weeds which otherwise compete with crop plants for light and soil nutrients.

Two of the most suitable herbicides are glyphosate and glufosinate. Transgenic varieties of maize have now been developed to each of these herbicides. Benefits from herbicide-tolerant crops have been suggested for the farmer, the environment and the consumer.

The farmer is predicted to benefit for a number of reasons. First of all because glyphosate and glufosinate are less expensive to purchase and easier to apply than some alternatives. Secondly, inducing herbicide tolerance in a crop may increase a farmer's flexibility because it may mean that an extra herbicide is available. Any existing herbicides can still be used because the fact that a crop has been genetically engineered to be tolerant to a particular herbicide does not mean that that herbicide – or indeed any herbicide – has to be used. Thirdly, herbicides such as glyphosate are particularly effective. Their use should therefore lead to a greater range of weeds being controlled, resulting in higher crop yields.

Benefits to the environment have been predicted because the herbicides to which crop tolerance is being developed generally break down faster to non-toxic products in the soil and are less likely to leach into ground water. In addition, they are often active in smaller amounts meaning that the total mass of chemicals applied to a crop should be less. Finally, some of these herbicides reduce the need for pre-emergence application (when the herbicide is applied before the seed germinates). The advantage of this is that post-emergence application can reduce the risk of erosion of fragile soils.

Benefits to the consumer are envisaged on two fronts. First of all, the technology, through increasing crop yields and requiring farmers to spend less on herbicides, should decrease food prices. Secondly, the technology should lead to lower rates of herbicide application, the use of less toxic chemicals and a decreased risk to domestic water supplies.

Are genetically modified foods safe for humans?

One potential hazard that has been suggested for genetically engineered foods, such as those made from maize, is that they may not be safe. It can be argued that nothing in life is totally safe. Nor is it the case that safety is necessarily to be equated with goodness. It is safer to read about the prevention of evil than to prevent it. However, societies rightly forbid many actions because it is considered that the likelihood of their being unsafe is too great for any benefits they might bring. So the question is 'Are genetically modified foods safe enough?'.

Most experts agree that the various transgenic maizes and other crops currently on the market are safe for animal and human consumption, being neither toxic or allergenic. That is why they have been approved by a large number of regulatory authorities in the EU, USA, Japan, Canada and elsewhere. However, some transgenic maizes have an antibiotic-resistant marker gene. The reason for this is that it makes it easier in the laboratory to see whether the genetic engineering has worked. The possibility has been raised that when large amounts of foods containing these antibiotic-resistant marker genes begin to be consumed (whether by farm animals or humans), the gene might move to disease-causing micro-organisms in the gut and so make them resistant to the antibiotic too.

Most experts suspect that the chances of this happening are not great. Nevertheless, the existence of a finite risk has slowed the regulatory approval of transgenic maize, notably in France. In the long run, the most likely solution is for companies to use other, less controversial markers. Technically this is possible, though less easy. In addition, there are some leading geneticists, such as Dr Arpad Pusztai, formerly at the Rowett Research Institute in Aberdeen, and Dr Mae-wan Ho of the Open University, who are worried that genetically modified foods are not safe.

Testing foods to see if they are safe is not easy. You can feed them to laboratory animals but then these animals don't live as long as humans do and such tests can't pick up small effects of the food on intelligence or personality. You can feed new foods to human volunteers. This is useful for identifying certain allergies but such feeding trials rarely last longer than weeks or a few months. Another approach is to carry out 'post-market surveillance'. This means that once a novel food has been introduced onto the market, you put in place some sort of monitoring system to try and pick up any harmful consequences, even if these are rare.

Questions

5 Explain why herbicides are so widely used by farmers in Europe and the USA.

6 Suggest reasons why farmers in African countries make less use of herbicides than farmers in Europe or the USA.

7 Draw up a table to show possible benefits of herbicide-tolerant maize to

a farmers,

b the environment,

c consumers.

Question

8 With reference to the testing of foods for safety, explain how observations and experiments can be used in science to rule out alternative explanations, with the aim of reaching a single, agreed explanation.

In October 1999, the *Lancet*, one of the world's top medical journals, published a controversial paper on genetically modified foods. The paper was authored by Dr Arpad Pusztai and described the effects of genetically engineered potatoes and other diets on the health of laboratory rats. Dr Richard Horton, the editor of the *Lancet*, defended his decision to publish the paper as being 'absolutely the right thing to do' despite the fact that the experiments had been reviewed by The Royal Society – Britain's top science research body – and condemned as flawed.

Discuss the decision to publish such controversial findings in the light of the scientific convention that reported findings and explanations must withstand critical scrutiny by other scientists before they are accepted as scientific knowledge.

Are genetically modified crops good or bad for the environment?

Genetically modified crops may turn out to be good for the environment. We reviewed above how transgenic maize might benefit the environment. However, it is possible that genetically modified crops may prove harmful for the environment. One possibility is that transgenic crops may be more likely to invade and then damage natural habitats. The likelihood of this will probably depend on the crop concerned. It is very difficult to imagine transgenic maize invading natural habitats in Europe. Maize is quite a difficult crop to grow and there are no examples of conventional maize invading or damaging natural European habitats. However, some crops, such as oilseed rape, can, to a certain extent, invade natural habitats (Figure 7.9). It is possible that genetically engineered varieties of such crops might prove more difficult to control than conventional varieties.

Figure 7.9

A researcher studying genetically modified rape in a field near King's Lynn, England. Some crops regularly escape from agricultural fields and may subsequently survive and reproduce in disturbed or semi-natural habitats. Oilseed rape can escape from a field and grow on roadside verges or in hedgerows.

Another potential problem with genetically modified crops is that genes from them might escape, via pollen, to weeds, wild relatives of the crop or other plants. A third potential problem is that transgenic crops might lead to a reduction in biodiversity. This might result from the fact that such crops may be toxic to any insects that eat them. Such insects may in turn be eaten by insectivorous birds. Recent decades have already seen large decreases in the numbers of many bird species, such as skylarks, that used to be abundant on agricultural land. There have also been large decreases in the numbers of certain arable plant species.

Assessing such environmental risks is not straightforward. This is partly because of the interdependence of species. In nature, changes to one species may have unintended and undesirable effects on other species. Scientists have tried to develop various models to predict the ecological consequences of genetic engineering. For example, a small plant with a very short life cycle called *Arabidopsis* (Figure 7.10) has been used as a model of larger crop plants. Experiments with *Arabidopsis* have shown that transgenic *Arabidopsis* is 20 times more

Figure 7.10

The small plant Arabidopsis thaliana *is used as a model for crop plants in genetic research.*

likely to cross-pollinate ordinary non-genetically engineered *Arabidopsis* plants than is non-genetically engineered *Arabidopsis*. This finding came as a complete surprise to the industry.

Other scientists have developed mathematical models to try to predict the population growth of and evolutionary change in plants with such escaped genes. A rather unkind, but possibly accurate, judgement on these models is that they tell us more about the brains of the people who devise them than about plant ecology. Indeed, the history of ecology has largely been a history of organisms doing things that biologists confidently believed they couldn't.

For example, after the explosion at the Chernobyl nuclear power plant in northern Ukraine on 26 April 1986, restrictions were placed on the sale of sheep in certain parts of Britain because the grass they were eating was contaminated by caesium-137. It was predicted by government scientists and others that these restrictions would be needed for about 3 weeks. In 2000, 14 years after the explosion, there were still some farms in Britain not allowed to sell their sheep because their meat contained too high a level of radioactivity. *Ref to Chapter 13*

The interdependence of species

No species exists in isolation. All living species affect other species. Feeding relationships between species are particularly important. For example, all animals survive by eating plants or other animals. These feeding relationships can be represented in food webs which show the flow of energy and nutrients from plants up through the other organisms in an area.

Another type of relationship between species is competition. In any given environment, there is competition between several species for the materials they require to live and reproduce.

As a result of such complex patterns of interdependence, changes which affect one species can have extensive knock-on effects on other species. Some disruptions lead to changes which have little effect on the other species. Other disruptions can lead to irreversible, large-scale effects.

Other current and possible future uses of genetically engineered plants and animals

Currently, the main use of genetically modified plants is, as described above, to provide herbicide-resistant and pest-resistant crop varieties. Looking to the future, genetically modified plants may be used to reduce human diseases. For example, a new strain of rice has been bred, using genetic engineering, to produce vitamin A and be high in iron. It is hoped that this will help literally hundreds of millions of people who are short of vitamin A or iron.

Plants are also being genetically engineered to produce vaccines. The idea is that people would develop immunity to diseases simply by eating certain plants rather than by having to go to a doctor or nurse for an injection. In parts of the world where refrigerators (needed to store conventional vaccines) are in short supply, this might lead to more people being immunised against diseases.

Questions

9 Explain precisely why crops genetically engineered to be poisonous to any insects that tried to eat them might lead to a decrease in insectivorous birds and mammals, such as skylarks and harvest mice, and to a decrease in carnivorous birds, such as sparrowhawks and owls, that eat small birds and mammals.

10 Suppose that the introduction of genetically modified crops leads to fewer pesticides being used on farms. Explain how this might be beneficial for the environment.

11 Make a list of possible ways that GM crops might harm the environment.

Discussion point

It is sometimes claimed that genetic engineering is just an extension of traditional agricultural technology. How valid do you consider this argument to be? Think about similarities and differences between traditional agriculture and genetic engineering with respect to such things as the speed of the technology, its scope, its precision, its effects on the environment, its safety and its public acceptability.

The main current use of genetically modified animals is as models of human diseases. Genetic engineering is used to breed a strain of laboratory mouse or rat that mimics a particular disease. For example, mice strains have been bred to have such conditions as various cancers, high blood pressure, cystic fibrosis or sickle cell anaemia. These mice are then used to try out new treatments for the conditions. The hope is that in this way new treatments will be found for humans suffering from these conditions.

Controversy exists, though, as to precisely how valuable or necessary the use of genetically engineered animals is in medical research. Some people, including the majority of medical researchers, maintain that their use will be of great value. Others believe that improvements in alternative approaches (including cell culture, tissue culture and computer modelling) mean that animals are no longer needed for such work.

People who campaign against the use of animals in experiments are concerned that the fall over the last 20 years in the number of animal experiments in the UK (Figure 7.11) may soon be reversed. The late 1990s saw a dramatic increase in the number of genetically engineered animals bred and experimented on.

Figure 7.11

The number of animal experiments in the UK fell throughout the 1980s and 1990s. However, the advent of genetic engineering may reverse this trend.

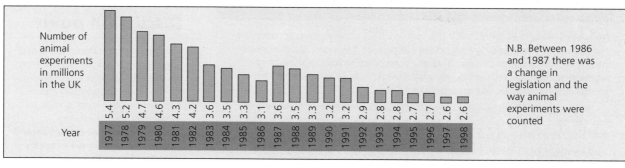

Number of animal experiments in millions in the UK

N.B. Between 1986 and 1987 there was a change in legislation and the way animal experiments were counted

Year	1977	1978	1979	1980	1981	1982	1983	1984	1985	1986	1987	1988	1989	1990	1991	1992	1993	1994	1995	1996	1997	1998
	5.4	5.2	4.7	4.6	4.3	4.2	3.6	3.5	3.3	3.1	3.6	3.5	3.3	3.2	3.2	2.9	2.8	2.8	2.7	2.7	2.6	2.6

Pigs for human transplants

Another purpose for which animals are being genetically engineered is to serve as human transplant sources. The basic idea is that pigs will be genetically engineered so that their internal organs carry some human proteins on their surfaces. The presence of these proteins should hopefully allow doctors to use the pigs' organs as transplant organs for humans. (This would be called 'xenotransplantation'.) This is because the proteins essentially fool our immune systems into thinking that instead of a patient receiving a pig's heart, kidney or liver, they are getting a human one instead.

Why use pigs in this way? The answer is that there simply aren't enough human organs to go around. Literally thousands of people every year die waiting for a transplant that never arrives. Each year the shortage is getting worse. In part, and somewhat ironically, this is because fewer people die in road accidents than was the case ten years ago. Such crash victims, though dead themselves, often help save other people's lives by providing organs for transplants. Pigs are the animals of first choice because their hearts are about the same size as ours and because their physiology is surprisingly similar to ours in some respects.

Question

12 Suggest ways in which the shortage of organs for transplants might be tackled without using genetic engineering. Is it likely that the methods you suggest will ever meet the demand for transplants?

Genetically engineered salmon

Research is underway to genetically engineer a number of fish species, including Atlantic salmon, for faster growth, tolerance of low temperatures or resistance to certain diseases. A number of benefits for this can be imagined. In particular, jobs would be created and salmon, a desirable and nutritious food, should become more abundant and cheaper.

Possible hazards of future examples of genetic engineering

We reviewed above the possible harms to human safety or the environment that might result from genetically modified crops. Might the use of genetically modified animals also prove hazardous? The short answer is 'yes'.

Pigs for human transplants

A number of national and international ethical committees have looked at the safety of genetically engineering pigs for human transplants. We know that pigs carry what are called porcine endogenous retroviruses ('PERVS'). In the light of BSE and AIDS it is unsurprising, and encouraging, that there is tremendous hesitancy in allowing any scientific/technological procedure to go ahead that might lead to new human infections.

The current position in the UK is that if (and it is a big 'if') xeno-transplants are allowed, the safety requirements will be stringent. The United Kingdom Xenotransplantation Interim Regulatory Authority is proposing that anyone receiving a xenotransplant must agree for the rest of his or her life to a whole set of conditions including: use of barrier contraception; refraining from pregnancy/fathering a child; and allowing the relevant Health Authorities to be notified when moving abroad. In addition, all household members and sexual partners will need to be seen before the operation to ensure they are informed about the possible risks, how to minimise them and to have blood samples taken for indefinite storage.

Genetically engineered salmon

Consider the use of genetically modified salmon in fish farms (Figure 7.12). The salmon would almost certainly escape. Fish farmers may give assurances that their animals are safely contained but the reality is that escapes of farmed Atlantic salmon from commercial fish farms occur frequently off the coasts of Norway, the UK and Canada.

> **Discussion point**
>
> If you were seriously ill and needed a transplant would you consider agreeing to receiving an organ from a genetically modified pig bearing in mind the conditions likely to be imposed on the recipients of a xenotransplant?

Figure 7.12

A salmon farm in the western isles of Scotland. Fish farms around the world have led to cheaper fish. Should fish be genetically modified to grow even faster?

During the 1990s, more than 100 000 farmed Atlantic salmon escaped from their net cages along the coast of British Columbia alone.

How serious would escapes of genetically engineered salmon be? That is a more difficult question to answer. Suppose first that the escaped fish are unable to breed with wild fish stocks. They might still spread diseases to other fish. Fish farms have been plagued by various diseases, notoriously lice and a virus that causes salmon anaemia. Indeed, diseases may be spread even if the fish are contained within the cages visible in Figure 7.14 as these only contain the fish, allowing water to enter and leave freely.

Escaped fish, even if unable to breed with wild fish, might also cause problems through competition or predation. As countless examples of the introduction of non-native species have shown, it is extremely difficult to predict the ecological consequences of the release of a species into a foreign area. Some escaped farmed Atlantic salmon (not genetically engineered) have successfully bred along the coast of British Columbia giving rise to concerns that they will compete with slower growing native Pacific salmon. The native salmon are already in decline due to disturbance to their habitat resulting from logging and other causes.

The chances of escaped farmed fish, whether genetically engineered or not, breeding with wild fish can be greatly reduced by various techniques which render fish infertile. However, these techniques are not 100% effective. If some fertile genetically engineered Atlantic salmon did escape into the wild, it is likely that the gene pool of the wild salmon would alter. The consequences of this are uncertain.

Finally, it is possible that escaped genetically engineered Atlantic salmon might hybridise with other fish species. For example, there is evidence of a certain amount of hybridisation between Atlantic salmon and brown trout. Again, the consequences of such hybridisation are uncertain.

Question

13 To what extent does salmon ecology illustrate the interdependence of species?

Ethical issues raised by genetic engineering

It is important to distinguish between technical issues (what can be done) and ethical issues (what ought to be done) when considering science and technology. Many people feel that genetic engineering is unethical. Consider the use of pigs for human transplants. True, there are major ethical arguments in favour of this if it will really lead to many human lives being saved. But there are also ethical arguments against, and people's views whether or not genetic engineering is acceptable may stem from more fundamental moral positions which they hold. For example, do we have the right to use pigs in this way? Not everyone accepts that non-human animals have rights. If you do, should that also mean that you believe it's wrong to have animals as pets – just as it would be wrong to have a human as a pet?

Aside from animal rights, there are welfare issues to do with the pigs being bred for human transplants. Companies involved in research on xenotransplantation maintain that their pigs are extremely well looked after. Indeed, in the UK, the pigs are actually looked after better than are pigs on most pig farms. Imutran, the major British company involved in the research, uses what is widely agreed to be a high quality animal welfare system to house its pigs. This comprises a warm, insulated bed and a cooler area for relaxing, feeding and drinking. It gives pigs a choice of environment and temperature and provides for social contact.

However, there is more to the welfare of the pigs than their housing. If and when clinical trials begin, it seems likely that some so-called germ-free animals will be needed. Such animals would be obtained by what is sometimes euphemistically called 'surgical derivation'. This means that shortly before birth, the entire uterus with the piglets would be removed (surgical hysterectomy) from the mother. The piglets would then be raised in isolation and in sterile conditions.

A more general ethical question about genetic engineering which applies to genetically engineered crops as well as animals is 'Is genetic engineering unnatural?' The simple answer is 'yes'. In nature you don't, for example, get human genes moving into pigs.

However, is 'natural' to be equated with 'good'? Presumably small-pox, earthquakes and death are natural whereas vaccines, laptops and AS courses in Science for Public Understanding aren't. In other words, there doesn't seem to be much of a relationship between what is 'natural' and what is good.

Aside from psychological reasons for the success of appeals to nature, one great advantage of nature is that it has been around for quite a while! Consciously or otherwise the thought may be 'Our ancestors successfully brought up their children, farmed and prepared their food in these ways so traditional approaches must be OK'. After all, and quite logically, one cannot be sure about the long-term consequences of any new technology (including genetic engineering), only of practices that have been around for a considerable time and so are now considered 'natural'.

Do people want genetic engineering?

Whether or not genetic engineering is ethically acceptable, do people want it? There is a widespread perception among consumers in Europe that foods containing products made from some genetically modified crops (notably soya and maize) are being forced on the general public. This is largely because since 1992 USA labelling policy has been based entirely on safety considerations rather than on enabling people to choose what sorts of products they wish to purchase.

The wave of public opinion against genetically modified foods that arose in Britain in 1999 illustrates the ways in which appropriate solutions to technological problems are influenced by a range of considerations. In the case of GM foods these considerations include technical feasibility (are antibiotic-resistance genes necessary?), economic cost (will GM foods be cheaper?), social impacts (will small farmers be forced out of business?), environmental impacts (will

Questions

14 Summarise ethical arguments both for and against the use of pigs for human transplants.

15 Can you work out to what extent your own views on genetic engineering are the result of what you have heard in the media, what you have been taught in school/college, the fundamental moral positions you hold or other factors?

16 Draw up a table of points for and against the 'it's unnatural' argument.

Figure 7.13

A Greenpeace protester demonstrating against genetically modified crops.

Question

17 Is it appropriate, or merely inevitable, for newspapers and other mass media such as television, to affect the ways we think about scientific issues? Explain your answer.

Figure 7.14

A surprisingly large number of committees and organisations are involved in the regulation of GM crops in the UK.

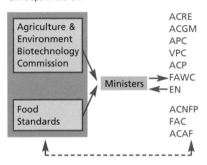

ACRE- Advisory Committee on Releases to the Environment
ACGM - Advisory Committee on Genetic Modification
APC - Animal Procedure Committee
VPC - Veterinary Products Committee
ACP - Advisory Committee on Pesticides
FAWC - Farm Animal Welfare Council
EN - English Nature
ACNFP - Advisory Committee on Novel Foods and Processes
FAC - Food Advisory Committee
ACAF - Advisory Committee on Animal Feedingstuffs

Question

18 How do you think people should be appointed to the government committees that decide whether or not genetically modified foods should be allowed to be sold in the UK?

GM crops be good or bad for wildlife?), ethical implications (is genetic engineering morally acceptable?), political issues (will approving GM crop plantings harm a government's standing in the opinion polls?) and religious attitudes (is genetic engineering 'playing God'?).

In the case of genetically modified foods, one of the most interesting features has been the role of the mass media. To the bemusement of many scientists, certain national newspapers have waged a campaign against the technology. Whatever your views about genetic engineering, the whole saga shows how mass media not only provide information but help set the agenda and influence public opinion. Nor is it only the mass media that can affect how people think about issues. Local pressure groups, non-government organisations such as Friends of the Earth and Greenpeace (Figure 7.13) and even famous individuals such as Prince Charles all affect public opinion about genetic engineering.

How is genetic engineering regulated?

It is useful to recognise the tensions between the rights of individuals and groups to do what they like and the need for society to regulate activities and practices for the common good.

In the case of the regulation of genetically modified crops in the UK, there are a considerable number of committees and other groups involved (Figure 7.14). Of central importance is who is appointed to these committees. For example, the ACNFP (Advisory Committee on Novel Foods and Processes) is the committee which recommends to government ministers whether new foods, including all genetically modified foods, should be approved. All but two of the members are scientists, several with links to the food industry.

Review Question

19 Some people have argued that genetically modified crops will be of no benefit to people in developing countries. Others have argued that genetically modified crops have the potential to help people in developing countries and that there is a moral duty for scientists in developed countries to help produce such crops.

Draw up a table of arguments:

a in favour of there being genetically modified crops in developing countries;

b against there being genetically modified crops in developing countries.

To get you started, here are some points for you to think about:

• GM crops may be developed which can be grown in saline soils or areas with low rainfall, thus enabling more people to be fed. Such crops may lead to greater food supply. They may also lead to an extension of agricultural land, reducing the area for wildlife.

• GM crops may be devised with particular nutritional and medicinal benefits. For example, in addition to the GM rice, high in vitamin A and iron, plants may be bred that are richer in proteins or don't need special treatment to make them less toxic.

• GM crops will cost more than conventional seeds but may have higher yields.

• GM crops may require inputs of fertiliser which will have to be bought.

Chapter 8

Evolution: understanding who we are

The issues

The theory of evolution has major implications for how we see ourselves. If most biologists are correct, humans and all of life on Earth shared a single common ancestor some three and a half thousand million years ago.

Yet theories about evolution have always been controversial and never more so than nowadays. The general public in a number of countries is deeply divided over whether evolution really has taken place. Many people doubt that it has and many Christians, Muslims and other religious believers maintain that the world is only some 10 000 or so years old. Other people happily combine their religious faith with an acceptance of the theory of evolution.

The science behind the issues

Scientists now know a tremendous amount about how individuals pass on genetic information to their offspring. They also know a great deal about how changes can take place to the genetic make-up of species. Such changes begin with mutations — alterations to an individual's genes. Most mutations are harmful and individuals with them are less likely to survive and reproduce. Occasionally, though, such mutations prove beneficial. The process by which species change over time as a result of the accumulation of beneficial mutations is known as natural selection.

Some biologists believe passionately that natural selection alone is sufficient to account for all of evolution. Other biologists believe equally strongly that natural selection alone is not enough.

What this tells us about science and society

The proposal of natural selection as a plausible mechanism for evolution was crucial to the acceptance of the wider concept of evolution itself.

Because evolution, if it occurs, takes so very long, testing theories about evolution is rather different from testing theories in most other scientific disciplines. There is a limit to how much designed experiments can tell us about evolution. Instead scientists have to be able to interpret evidence gathered in various ways. For example, fossils have to be studied and conclusions drawn from their structure and distribution (Figure 8.1). This means that a great deal of interpretation is required.

Figure 8.1

Rodolfo Coria dusting the teeth on the fossil jawbone of the largest carnivorous dinosaur discovered so far. He unearthed the 110-million-year-old bones of this dinosaur in 1993 at El Chocon, Argentina.

Figure 8.2

A portrait of Charles Darwin in his early 20s. Darwin is generally regarded as the most important biologist who has ever lived – though some historians of science disapprove of such rankings.

Question

1 To what extent does Darwin's work illustrate the point that people become more confident about a scientific explanation if the theory includes a plausible mechanism for the changes observed?

Charles Darwin's life and work

The person whose name is most remembered in connection with the theory of evolution is Charles Darwin (Figure 8.2). This is not because he was the first person to suggest that evolution had occurred. On the contrary, for over 2 000 years various people had argued that species had evolved from very different ancestors. However, Darwin is rightly famous for two reasons. First, he produced a mass of evidence that succeeded in convincing both other scientists and many members of the general public that evolution had occurred. Secondly, he provided a convincing theory as to how evolution had occurred – the theory of natural selection.

Darwin's life

Charles Darwin was born in 1809 while England was deep at war with France. His mother, Susannah, was then 44 years old. His father, Robert, was a very successful doctor. His family had high hopes that Charles too would be a doctor but as a medical student Charles soon realised that he couldn't face operations and the suffering of patients.

As a child, Charles, in the words of his biographers, Adrian Desmond and James Moore, 'was an inveterate collector and hoarder – shells, postal franks, birds' eggs, and minerals'. He also seems to have been quite an insecure child and for much of his life tried to win his father's affection in various ways. As a boy, for example, he once made off with his father's peaches and plums, then hid them so that he could 'discover' and report the find the next day. On another occasion, he invented an elaborate story intended to show how fond he was of telling the truth.

As an undergraduate, Charles' collecting instincts focused on insects. He became passionate about beetles and found nothing more thrilling than finding a new specimen. However, the question as to his future career remained. At one point it looked as though he might get ordained and become a vicar, but that idea petered out. And then, remarkably, at the age of 22 he was offered the chance to spend 2 years on a small ship, the HMS *Beagle*, that was to sail round the world making maps (Figure 8.3).

Historians today are still unclear why Charles was offered this post. Textbooks generally state that Darwin was the ship's naturalist. Yet the ship set sail with another man as the official naturalist. A different possibility is that Darwin, as a gentleman, was meant as a suitable dining companion for the ship's captain, Captain Robert Fitzroy. They must have made a remarkable pair. Fitzroy was a man of great temper and had a firm belief throughout his life in the literal truth of the bible and the impossibility of evolution. Later in life he made an important contribution to the science of meteorology.

Fortunately, Darwin seems to have been remarkably good at getting on with people. During the voyage he managed to maintain pretty good relationships with Fitzroy. However, the five years that the voyage eventually took were long enough for Darwin to return home convinced that evolution was a fact.

How did Darwin get his evidence for evolution?

During his travels Darwin gradually became convinced that species were not fixed: that they had evolved over the course of extremely long periods of time from simpler ancestors. Part of the reason he changed his mind was that he spent a lot of his time collecting fossils. Some of these fossils resembled species alive in Darwin's time but were clearly different from them. Evidently such fossils were the remains of extinct species.

Another reason came from the study of living species. Darwin was especially impressed by the Galapagos Islands (Figure 8.4). These are tropical islands about 900 km west of Ecuador, off the coast of South America. The *Beagle* spent a month among the islands and the islands provided Darwin with some of his best evidence for evolution.

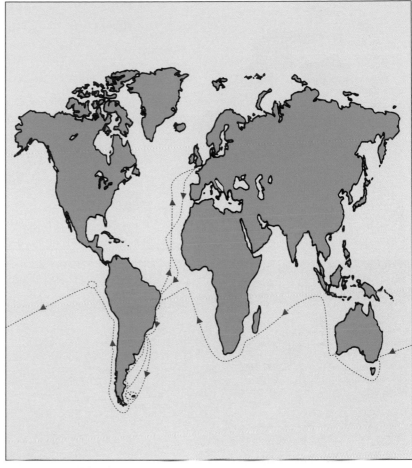

The most striking thing about the organisms on the Galapagos Islands is that few of them are found on the mainland of South America yet all of them resemble organisms found there. For example, there are 13 species of finch. All are unique to the islands but resemble finches found on the mainland. And then there are the tortoises. The Spanish for tortoises is *galápagos* and so it is these animals that have given their name to the islands. These tortoises are big (Figure 8.5). They can reach a metre in height and have a mass of up to 225 kg.

What really intrigued Darwin about the tortoises, rather than their impressive size, was a chance remark made to him one evening at dinner by a certain Mr Lawson. Mr Lawson was vice-governor of the Galapagos Islands and happened to say to Darwin that he could tell by looking at a tortoise from which island it had come. This set Darwin thinking. How could this be? Was each sort of tortoise created independently on each island? Or did all the tortoises share a common ancestor and it was simply that on the different islands, separated from one another by deep water, the tortoises had evolved over time so that they were now recognisably different?

But if tortoises could evolve thus, perhaps, Darwin mused to himself after he had left the islands, the same was the case with the 13 species of finch. Unfortunately, Darwin hadn't thought of this possibility while

Key term

Fossils are the remains of organisms that have died. In certain circumstances fossils may faithfully preserve details of the structures of organisms that lived long ago.

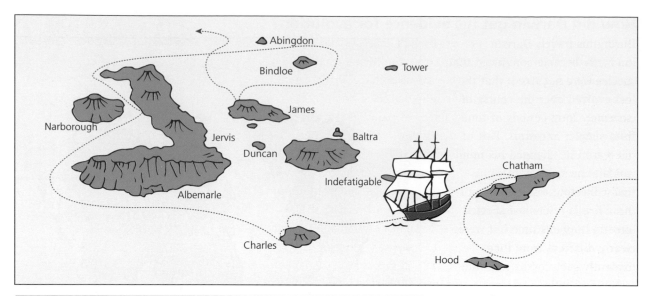

Figure 8.4

Map showing the voyage of the Beagle through the Galapagos Islands off the west coast of South America.

Figure 8.5

An 1886 illustration of the giant tortoises of the Galapagos Islands. These tortoises helped convince Darwin that evolution was a fact.

still on the islands so he hadn't bothered when collecting bird specimens to label from which island each one came.

Nevertheless, writing in one of the many notebooks he kept during his life and in which he wrote his scientific ideas as they occurred to him, he suggested that perhaps the finches were all descended from one species.

Questions

2 Explain why Darwin thought that his observations of creatures on the Galapagos Islands could be explained by evolution.

3 Why didn't Darwin label the Galapagos finches with the names of the individual islands from which he had collected them? Do you think his failure to do this shows that he wasn't a good scientist?

Natural selection

Back home after his long voyage, Darwin tried to come up with a mechanism that could account for evolution. For some scientists, there is a sudden point when a new theory takes shape in their mind. For others, the process takes much longer. Darwin's theory of natural selection seems to have taken a long time coming but was greatly helped by his reading an essay written in 1798 by the English clergyman, Thomas Malthus.

Malthus argued that while the human population increases geometrically, the available food supply only grows arithmetically (Figure 8.6). As a result, Malthus predicted, population growth will always tend to outstrip the available food supply. Even at the time Malthus' argument was controversial and it is difficult to see why food supply can only increase arithmetically. However, Darwin saw that Malthus' argument could be applied not just to humans but to all species. The key idea he got from Malthus is that there is a struggle for existence. Malthus thought of this struggle as the struggle for food. Darwin realised that it could take various forms.

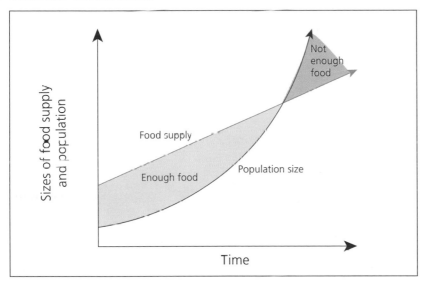

Eventually Darwin came up with four basic ideas:

- Individuals within a species differ from one another.
- Offspring generally resemble their parents.
- In all species more offspring are born than can survive to adulthood and reproduce.
- Some individuals are better suited to their environment than others.

As a result, Darwin reached two conclusions:

- Those individuals that do survive to reproduce pass onto their offspring the characteristics that enabled them to survive.
- Over time, a group of individuals that once belonged to one species may give rise to two different groups that are sufficiently distinct to belong to separate species.

Key terms

An increase is said to be **geometric** if its rate of doubling is constant. For example, if a population of weeds in a field doubles every four months, its increase is geometric.

An increase is said to be **arithmetic** if the absolute increase in numbers is steady over time. For example, if a population of weeds in a field increases by 1000 individuals every year, its increase is arithmetic.

A **species** is a group of organisms that have the capacity to breed with one another. So, for example, orang-utans and gorillas belong to different species.

Figure 8.6

A plot of food supply and population size showing food supply growing arithmetically while population is growing geometrically. Malthus maintained that in the long run population growth would always outrun food supply. Malthus' reasoning helped both Darwin and Wallace to come up with the theory of natural selection

Questions

4 Explain why natural selection would not work if:

a individuals within a species were all the same;

b offspring did not resemble their parents;

c all offspring survived to reproduce;

d individuals were all equally suited to their environment.

5 Distinguish between the theory of evolution and the theory of evolution by natural selection.

6 Identify features of Darwin's theory of evolution by natural selection which required conjecture and creative imagination.

Artificial selection as evidence for natural selection

These four ideas and the two conclusions that follow from them make up the theory of evolution by natural selection. Darwin spent years thinking about this theory and obtaining evidence for it. Some of the best evidence came from what is known as 'artificial selection'. Artificial selection is selection not by nature but by humans. Darwin spent many hours talking to people who bred farm animals and varieties of pigeons. He was particularly struck by one pigeon breeder's claim that he could produce a new variety within only a few years (Figure 8.7). Darwin realised that if artificial selection could produce a distinct new variety within a few years, perhaps nature could produce entire new species given enough time.

Typically for Darwin, his enthusiasm for pigeons, once ignited, knew no bounds. He built a pigeon house in his garden, bought various breeds of pigeons – with such names as 'fantails', 'pouters' and 'almond tumblers' – and set to work breeding them, observing them and measuring them. To his delight he found surprisingly large differences between the various varieties. And not just external differences. Even their red blood cells were differently shaped. To Darwin's fascination he found that the common pigeon held within its various varieties enough variation for biologists to classify them as belonging to fifteen different species, had they been wild birds.

Figure 8.7

Breeds of pigeons exhibited in 1864 by one of the London pigeon clubs to which Darwin belonged. All pigeons are descended from the rock dove, as is the ordinary pigeon found in towns. The great diversity of pigeon breeds is due to artificial selection. This engraving appeared in The Illustrated London News *in January, 1864.*

Question

7 To what extent was Darwin using artificial selection as a model for natural selection? Explain the similarities and differences between natural selection and artificial selection.

Evolution before Darwin

Darwin was not the first person to suggest that species had evolved over time. Indeed, the idea had been suggested by the Ancient Greeks more than 2000 years before Darwin was born. Nor was Darwin the first person to come up with a coherent theory for evolution. 'Coherent' comes from the verb 'to cohere' that is, 'to hold together'. A theory is coherent if it makes logical sense; that is, if it is internally consistent and provides a logically acceptable explanation for what it is intended to explain. This is not the same as a theory being correct or true.

A Frenchman called Lamarck produced a coherent theory for evolution. It just happens to be wrong, though.

Darwin published his book *The Origin of Species by Means of Natural Selection or The Preservation of Favoured Races in the Struggle for Life* in 1859. (The Victorians liked long titles for their books.) Exactly 50 years earlier Jean-Baptiste de Lamarck had put forward his own, very different theory of evolution.

Lamarck suggested that any helpful characteristics that an individual acquires during its life can be handed on to its offspring. For example, a sprinter who had developed the ability to run very fast would be expected to pass on this ability to any children she had. Lamarck's theory therefore relies on the transmission of acquired characteristics.

As everybody knows, giraffes have long necks (Figure 8.8). But how have these long necks evolved? Both Lamarck and Darwin would presume that today's giraffes have evolved from ancestors with shorter necks. However, they would explain the presence of long necks in today's giraffes very differently.

Lamarck's theory sounds a bit like one of Rudyard Kipling's *Just So Stories* – the one in which the elephant gets its trunk by having its nose pulled by a crocodile. A Lamarckian explanation would rely on the idea that when food was in short supply, giraffes would try to reach food at the tops of tall trees. In doing so, they would stretch their necks. This characteristic – i.e. long necks – is then passed on to any baby giraffes they have.

A Darwinian explanation would be rather different. It would rely on the idea that there are always more giraffes than there is food for. As a result, those giraffes that just happen to have taller than average necks would be most likely to survive and produce baby giraffes. The characteristic of having long necks would then be passed onto the next generation.

Figure 8.8

Lamarck and Darwin came up with alternative theories to explain why giraffes have long necks.

Wallace's theory of natural selection

On 18 June 1858 Darwin received a 12-page letter out of the blue. He was stunned. The letter came from a relatively unknown naturalist called Alfred Wallace and exactly described Darwin's own theory of natural selection. Looking back, we can see that this isn't the amazing coincidence it might seem to be. After all, the theory of natural selection is a logical theory and both Darwin and Wallace had read Malthus' arguments about there never being enough food to keep up with the growth in population. It's a bit like some theories in mathematics which are now known to have been independently thought of in India, in China and in Europe. In this way science is different from literature, music and art. Only Shakespeare wrote *Hamlet*; only Beethoven composed the *Pastoral Symphony* and only Leonardo da Vinci painted the *Mona Lisa*. Both Darwin and Wallace thought of the theory of natural selection while spending years away from home studying the amazing diversity of life in the tropics.

Wallace came from a much less privileged background than Darwin. He had left school at the age of 14 because of his family's financial difficulties and became a teacher. A stroke of good fortune led to him spending 4 years on an expedition to the Amazon. Despite appalling conditions he collected nearly 15 000 animal species, some 8 000 of which were new to science.

Then he had a terrible misfortune. On the way home the ship caught fire and all 15 000 species were lost together with most of Wallace's notes. Despite this, Wallace set off in 1854 for an 8-year expedition to Asia. It was there, as he lay ill with fever in 1858 that he wondered:

"Why do some die and some live? ... from the effects of disease, the most healthy escape; from enemies, the strongest, the swiftest, or the most cunning; from famine, the best hunters or those with the best digestion."

Recovering from the fever, Wallace wrote up his theory and sent it to Darwin. Together they published a paper on the theory and for that reason it really should be known as the Darwin-Wallace theory of natural selection.

How long does evolution take?

Breeds of domestic species, such as dogs, sheep and domestic pigeons, differ greatly from the breeds of just a century ago. Yet in the wild, species seem remarkably constant. Drawings, paintings and carvings of animals from the Egyptian tombs show no recognisable differences from today's species. Yet such representations were made almost 4000 years ago. Nor do prehistoric paintings of animals seem different from the same species today (Figure 8.9). Yet some of these paintings are thought to have been painted 20 000 or 30 000 years ago.

So just how long does evolution take? Most evolutionary biologists think that a typical mammal species takes about two million years to evolve into another mammal species. And the two species would look so similar that you would have to be quite an expert to tell them apart. So evolution is a very slow process. Darwin didn't know how long it took species to evolve but he appreciated that his theory only made sense if the world was very old. Fortunately for Darwin, a growing number of geologists also

Questions

10 List the similarities between Darwin and Wallace that may have led to them independently thinking of the theory of evolution by natural selection.

11 Why do you think that today it is Darwin rather than Wallace that most people associate with the theory of evolution?

Figure 8.9

Prehistoric painting of big-horn sheep carved into the rock by Native Americans around 12 000 years ago. The rock face is now part of the Capital Reef National Park in Utah, USA. Such paintings suggest that evolution, if it has occurred, has taken place only very slowly.

believed that the world was very old. Just how old no one knew but certainly many hundreds of millions of years old, possibly thousands of millions of years old.

Nowadays the dating of rocks through the principle of radioactive decay suggests that the Earth itself is about four and a half billion years old – i.e. four and a half thousand million years old. The first fossils have been dated at some three and a half billion years old. Not surprisingly, such fossils are very simple. They consist of organisms that have just a single cell. Multicellular organisms don't appear until about seven hundred millions years ago (Figure 8.10).

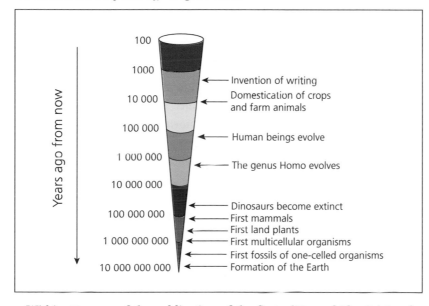

Figure 8.10

Timechart showing the main events in the history of life on Earth. Note that each unit on the scale is ten times larger than the one before. This is a logarithmic scale.

Within 10 years of the publication of the first edition of *The Origin of Species*, Darwin was shaken by the pronouncements of the physicist William Thomson. Thomson calculated that the Earth was only 100 million years old. One hundred million years may sound a long time but Darwin was greatly influenced by the geologist Charles Lyell who had argued that the Earth was very old. In 1859 Darwin had hazarded a guess that the reign of the dinosaurs had been 300 million years ago which would mean that the Earth itself must be many times older than that.

Darwin found it impossible to believe that 100 million years gave enough time either for evolution to occur or for all the thousands of feet of sediments visible in the fossil record to have been laid down. Nevertheless, Darwin accepted Thomson's calculations which were based on the rate at which the Earth was cooling down. He therefore spent the winter of 1868–69 making lots of alterations to the latest edition of *The Origin of Species*. In this fifth edition, which appeared in 1869, Darwin not only claimed that the environment caused a greater number of useful variations to appear but he also embraced Lamarck's idea of the inheritance of acquired characteristics. So certain was Darwin of the evidence for evolution that he was prepared to allow Lamarck's theory for its mechanism to sit alongside his and Wallace's theory of natural selection.

Questions

12 Why was Darwin worried by the idea that the Earth might be much younger than he had thought?

13 Do you feel Darwin was right to believe that 100 million years was enough time for evolution to have occurred or should he have abandoned his theory of evolution in the light of Thomson's conclusions? Comment on the facts (a) that he wrote his fifth edition of *The Origin of Species* to suggest that 100 million years was long enough and (b) that he introduced Lamarck's theory alongside his own theory of natural selection.

Decades later, long after Darwin had died, physicists realised that while Thomson's mathematics was correct, his calculations were based on a faulty model. It turns out that the Earth is kept warm partly by radioactivity, something Thomson knew nothing about as radioactivity wasn't discovered until 1896. Darwin needn't have worried about the age of the Earth.

Reactions to the theory of evolution

By 1838 Darwin was convinced that his theory of natural selection explained evolution. Yet he didn't publish his results for over 20 years. Indeed, he might never have published them had not the fateful letter from Wallace arrived. Why was Darwin so reticent? One possibility is that he was simply being cautious, preferring to gather enough evidence to support his theory. And certainly, the sheer size of *The Origin of Species* – the book contains some 500 pages of prose with just one diagram – and weight of evidence it revealed helped his work to gain acceptance. By 1859 he was also a well known and very widely respected scientist. He had an excellent scientific reputation and many of his previously published ideas had withstood critical scrutiny by other scientists.

But there is another possibility and that is that he was afraid of public reaction (Figure 8.11). It is difficult for us, nowadays, to imagine how revolutionary his ideas sounded to Victorian England. At that time most people accepted the literal truth of the Christian and Jewish scriptures. These imply that the Earth could not be more than about 6 000 years old. Yet Darwin needed the Earth to be many hundreds of millions of years old.

Perhaps more importantly, the Bible tells how all humans are descended from Adam, a perfect, sinless immortal created by God in the image of God out of the dust of the Earth. This is a far cry from the Darwinian vision which has all of us descended via a near-endless succession of intermediaries back through primitive mammals, through invertebrates looking rather like present-day earthworms, through single-celled creatures barely visible without a microscope, back to a primordial soup.

Interestingly, some of the theologians of Darwin's day were quicker to accept his views than were many scientists. Nevertheless, some of the clergy of Darwin's day were vehemently opposed to Darwin's views. There were various reasons for this. Some clergy felt that the theory of evolution was treading on their area of expertise; others felt that if evolution had occurred, there could be no valid base to morality and people would be free to behave 'like animals'.

To this day, the theory of evolution remains controversial, especially for many religious believers (Figure 8.12). In 1999 the Oklahoma State Textbook Committee in the USA produced a statement which defined evolution as 'unproven' because 'no one was present when life first appeared on Earth'. The statement went on to say that 'evolution should be considered a theory, not fact'. In the same year the Kansas State Board of Education voted to remove evolution from its science curriculum. As a

Question

14 Do you think that Darwin's theory would have been more or less likely to have been well received if he had published it immediate after returning from his journey on the HMS *Beagle*? Explain your answer.

Figure 8.11

Once Darwin's theory of evolution by natural selection was published it was widely read and excited great controversy. This cartoon, published in the London Sketch Book in 1874, shows an ape-like Darwin holding a mirror up to an ape to show how alike the pair of them are.

Bill Day Detroit Free Press

NOBEL SCIENTISTS DISCOVER,
THE MISSING LINK---

EARTH IS ONLY 10,000 YEARS OLD!

AUSTRALOPITHECUS AFRICANUS HOMO ERECTUS CREATIONIST NEANDERTHAL MODERN MAN

Figure 8.12

Creationists believe that the Earth is at most 10 000 years old and that God created all species in much the form that they exist today. Nowadays creationists are often mocked in newspapers and magazines just as Darwin was in his time. .

Questions

15 Explain the reluctance of many people both in Darwin's time and today to accept the theory of evolution.

16 Do you feel that more scientific evidence for evolution would make creationists accept that it had occurred?

17 With reference to the theory of evolution, suggest how a person's expectations and their social interests and commitments can influence their interpretation of data.

18 Explain the difference between facts and theories. Use the distinctions you make to decide whether evolution and natural selection are facts or theories.

consequence, some publishers have begun to remove sections on fossils and evolution from school textbooks. One Kansas textbook for 12- and 13-year-olds originally started with a chapter on fossils, the origins of oil and gas deposits, and an account of the inland sea that is believed once to have covered the state. The new year 2000 version opens with the arrival of Native Americans, omitting all earlier events. The publisher said it wanted to improve the book's marketability. 'You don't want to offend any group in Kansas', a director of the publishing corporation stated.

The genetic basis for evolution

School textbooks usually state that Darwin knew nothing of the genetic basis for evolution. This is a mistake. We now know that the person most responsible for the science of genetics, Mendel, sent some of his findings to Darwin. The truth is, Darwin, along with every other scientist to whom Mendel sent his work, failed to appreciate its significance. Perhaps Mendel really was 'ahead of his time'.

Gregor Mendel was born in 1822 and became a monk (Figure 8.13). He entered a Moravian monastery (in what is now the Czech Republic). The monks were encouraged to study and Mendel went to Vienna University where he became interested in plant breeding. Over the course of a decade, from 1856 to 1866, Mendel bred, examined, described and counted over 28 000 garden peas. He discovered that the various characteristics of the peas – such as their height and the colour and shape of their seeds – were determined by what he called 'factors'. He also appreciated that each adult pea plant has each factor in pairs whereas the pollen and egg cells have only one copy of each factor. We now know Mendel's factors as genes. And Mendel was right. In most species there is just a single copy of each gene in the sex cells (sperm, pollen and eggs) whereas all other cells have two copies of each factor. Mendel was indeed the founder of modern genetics.

Figure 8.13

Gregor Mendel, the founder of modern genetics.

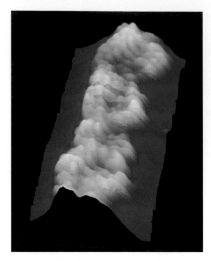

Figure 8.14

False-colour electron microscope image of a DNA double helix magnified about 2 million times. We now know that genetic mutations are tiny alterations in the structure of DNA. These changes are the building blocks for evolution.

Question

19 Explain what is meant by the word 'mutation' in terms of the chemical structure of DNA.

Figure 8.15

Evolutionary biologists think that all of today's life shared a single, common ancestor. The vertical axis shows time with most recent time at the top. The horizontal axis has no scale but is an indication of variation between species. The more lines there are on the diagram, the more species there were at the time. A point where two lines diverge from a single point indicates a time when one species split into two. A line that suddenly stops before today indicates that the species became extinct.

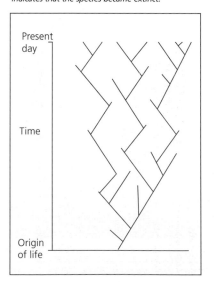

Mendel's work was rediscovered in 1899, 33 years after he had first published it. Soon the hunt was on to understand more about genes. Much of the work in the early decades of the 20th century was done on organisms that breed much more rapidly than we do – such as maize, mice and tiny fruit flies. Scientists soon appreciated that Darwin had been right when he saw inherited variation between individuals as lying at the heart of evolution. We now know that mutations are the source of this variation. *Ref to page 76*

Mutations are changes in the structure of an organism's genetic material. They can be caused by certain chemicals, including some found in cigarette smoke, by ionising radiation (ultraviolet rays, X-rays and gamma rays) and by radioactive emissions. If a mutated gene happens to be in a sex cell, copies of it can be handed onto offspring. Mutated genes may have no effect at all on the characteristics of an organism. Usually, though, they are harmful. Very occasionally they may be beneficial. *Ref to page 101*

The only good examples we have of beneficial mutations are of mutations that allow organisms to survive better as a result of human changes to the environment. For example, a great deal is known about how mutations have enabled many disease-causing bacteria to be resistant to various antibiotics. People who don't accept the theory of evolution conclude that there is no evidence that mutations ever help organisms in the natural environment. Evolutionary biologists maintain that it is simply that most of the recent changes to the environment have resulted from human actions.

The theory of evolution by natural selection

The fundamental principle behind the theory of evolution is that all the Earth's present day life forms have evolved from common ancestors and ultimately from self-replicating molecules which happened to develop under the conditions then prevailing on Earth.

Evolutionary biologists believe that all the species which now exist, and all those which have existed in the past, can therefore be linked in a single 'branching tree' structure (Figure 8.15). Many of the branches of this tree are 'dead ends'; most of the species which have existed at various times in the past have become extinct.

The mechanism which best explains evolution is natural selection. Chance mutations are the source of changes in individuals. Within a species, some individuals are better able to survive and reproduce in the environment in which they find themselves than are others. Over time changes accumulate and can lead to the evolution of new species. The enormous length of time required for this is consistent with our understanding of the geological time-scale.

Among the implications of the Darwin–Wallace theory of natural selection are that the process is not driven by any overall direction or aim. As a result, the path which evolution has taken could, with minor changes in circumstance, have been quite different.

Are humans at the top of the evolutionary tree?

We are used to seeing ourselves as the pinnacle of life. But are humans really at the top of the evolutionary tree? In one sense, aren't all surviving species equally successful? After all, if evolution is true, we, blue whales, emperor penguins, honey bees, oak trees, dandelions, edible mushrooms and every microscopic organism are all equally descendants from the original first living creatures that existed some four thousand million years ago (Figures 8.16 and 8.17). In what sense are humans, members of the species *Homo sapiens*, superior to all the other 10 million or so species still alive today?

The Darwinian view of life therefore causes us to re-examine ideas of progress and purpose. We may be more intelligent than other species but we lack the vision of an eagle, the hearing of a dog, the speed of a swift and the hardiness of many insects. What is progress? One view of evolution is that there is no upward inevitable ascent towards perfection. Instead, species merely respond to the environmental pressures they face. Humans have lost most of their ability to smell or to climb trees but they have gained the ability to imagine what life is like for other creatures and they have gained the ability to communicate with a far richer system of communication than any other species has.

And what is the purpose of life? The purpose of life for a spider is to produce baby spiders. The purpose of life for a salmon is to produce baby salmon. Is the purpose of life for humans to produce children? Or have we stepped beyond the confines of evolution so that we are able to determine for ourselves what our purpose is and how our life has meaning?

Does natural selection explain all of evolution?

Most biologists today accept that natural selection is the prime mover behind evolution. However, there is still controversy as to whether natural selection alone is responsible for evolution. Richard Dawkins has championed the idea of the selfish gene. Dawkins' view is that we should be thinking of evolution not at the level of species or even individuals but at the level of genes. Consider, for example, the way in which a worker honey bee never breeds and may even lose her life defending her hive. Clearly she herself benefits neither from her sterility nor from her altruistic (helping) tendency to protect the hive. But the genes in her live on via the other members of the hive.

Other biologists think that Dawkins overstates the importance of genes. One of the tasks of biology over the next 20 years or so is to find out whether genes alone are what drive evolution or whether there is more to the history of life than that.

Questions

20 Mendel's theory went far beyond observations and counting the characteristics of peas. He proposed a model of inheritance to explain his findings. Describe the model proposed and explain the differences between data which must be honestly observed and reported and a model, which may involve imagining things that cannot be observed directly.

21 Why does the general public nowadays accept the gene theory of inheritance when the leading scientists of Mendel's time failed to appreciate the significance of what he had found out?

Discussion point

Do you feel that chimpanzees and the other great apes should enjoy the same legal rights that we do? Defend your answer.

Figure 8.16

A chimpanzee drinking from a handful of leaves used as a sponge. Chimpanzees are probably our closest evolutionary relatives. We share about 99% of our evolutionary history with them and about 98.5% of our genes.

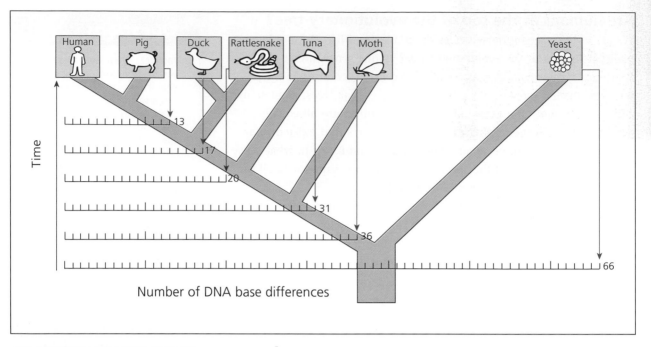

Human Pig Duck Rattlesnake Tuna Moth   Yeast

Time

13

17

20

31

36

66

Number of DNA base differences

Figure 8.17

Biologists can construct evolutionary trees based on the similarities between organisms. This evolutionary tree shows the number of base differences in the DNA structure of one molecule - the enzyme cytochrome C - found in almost all living organisms. The assumption is that the greater the number of differences between two species, the longer ago they separated from one another. Evolutionary trees are increasingly being constructed on the basis of similarities and differences between the DNA of organisms.

Question

22 List arguments

a in favour, and

b against the idea that human beings are superior to other species.

Discussion points

Do you find the Darwinian view of life demeaning for humans or ennobling for other species? Explain your answer.

To what extent do you think such human behaviours as selfishness and kindness can be explained by the selfish gene idea?

Discuss the extent to which we can draw any ethical conclusions from the fact that chimpanzees and humans share about 98.5% of their DNA in common. Try to come up with one set of arguments for this genetic similarity being of vital importance and another set of arguments for its not being very important. Present your arguments as opening speeches in a court case.

Review Questions

23 Produce a short book, of no more than 800 words but with illustrations, intended to tell 8-year-olds about the importance of Darwin's work. Bear in mind that some of your readers will come from families which accept the literal truth of the scriptures.

24 Use internet search engines and/or conventional libraries to find out more about continuing debates about the mechanism of evolution. You might find it helpful to consult www.nap.edu/readingroom/books/evolution98.

Chapter 9

Using fuels

The issues

The way that people live their lives depends to a great extent on fuels. The higher a person's standard of living, the more goods and services he or she can buy. Making these goods and providing these services requires fuels. In almost every country, people use more fuel today than they did in the past. People in wealthy countries use much more fuel (on average) than people in poorer countries (Figure 9.1). To reduce this inequality, we must either increase fuel use in the poorer countries (which would lead to problems of supply and to increased pollution), or reduce fuel use in the wealthier countries (which would have social, economic and political implications).

The science behind the issues

Fuels are valuable because they are concentrated energy sources. When we use a fuel, we transfer its stored energy to other places. Energy is not destroyed in the process. But it is spread out and becomes less concentrated – and so is less useful for doing anything more. Some processes use fuels more efficiently than others. Efficiency is a measure of how much of the energy goes where we want it to.

What this tells us about science and society

Our high standard of living is largely the result of technological developments, many of them based on scientific ideas. But the demands this makes on fuels, particularly fossil fuels, has rapidly depleted a precious finite resource, and leads to emissions of gases which damage the environment. The global inequalities in fuel use (and in standard of living) raise serious moral questions. While technical developments can contribute to tackling these problems, they cannot provide a complete solution. Economic, environmental and ethical considerations are often involved, and decisions involve balancing different considerations against each other.

Figure 9.1

(a) This life-style needs quite a lot of energy each year, USA 1950 (20 barrels of oil),
(b) This life-style needs even more energy, USA 2000 (30 barrels of oil);
(c) This life-style needs very little energy each year (mainly fire wood), Ghana 1990 (3 barrels of oil).

Fuels and energy

Fuels are useful because they are concentrated energy sources. When we burn a fuel, we release energy which we can then use to do other things (such as heating things up, making things move, and so on). Most of the energy for manufacturing industry, agriculture, transport and our homes comes from burning fossil fuels – coal, oil and natural gas. At present they are readily available, but they are a finite resource which will eventually run out. Some comes from nuclear fuel (enriched uranium and plutonium) which is used in nuclear power stations to generate electricity. In some parts of the world, wood or dried animal dung (biomass) is the main fuel.

But fuels are not our only energy sources. The Earth receives a great deal of energy directly from the sun, in the form of solar radiation. It can be used directly for heating, or to generate electricity using photo-voltaic (or solar) cells. The sun also drives the water cycle (which raises water into high dams for hydro-schemes), and causes wind and waves, which can also be used as energy sources.

Fossil fuels (the remains of ancient rainforests) take so long to produce that the current reserves are effectively all we have. They are non-renewable energy sources. On the other hand, wind, water and sunlight are renewable energy sources. So is wood, provided it is not used faster than trees can re-grow.

Figure 9.2

Fuels and other energy sources.

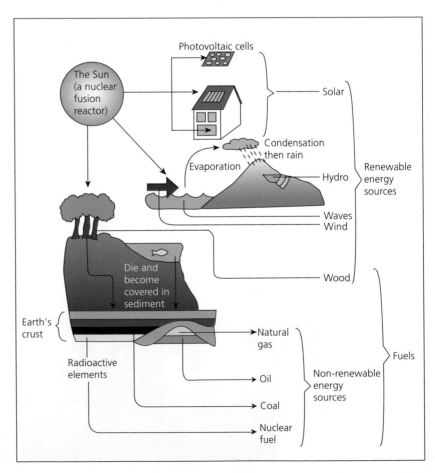

Primary and secondary energy sources

When fuels are burned in air, heat is released. Sometimes we use this directly, for example to heat our homes or cook food, or to make a car engine run. But we can also use it to make steam to drive turbines in power stations, which then turn generators to generate electricity. Electricity is then used in factories, homes and businesses to run a wide range of appliances (Figure 9.3). Electricity is said to be a **secondary energy source**. It has to be generated using a primary energy source (such as a fossil fuel, nuclear fuel or one of the renewable energy sources). Ref to Chapter 10

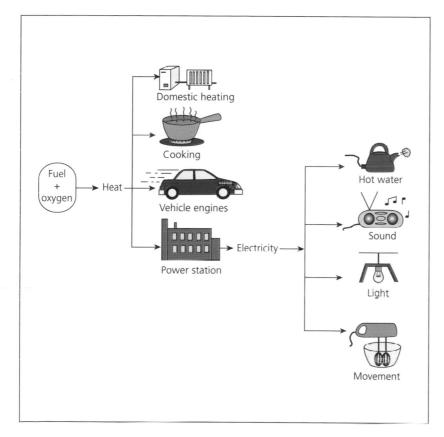

Figure 9.3

Different pathways from primary energy source to end use.

Key term

The scientific unit for measuring amounts of energy is the **joule** (J). This is the unit used in Figure 9.4.

Question

2 Look at the bar graph in Figure 9.4.

a Why is the bar for primary energy sources larger than the one for energy sources at point of use?

   Ref to Chapter 10

b Why are renewable energy sources not shown in the first bar?

c Which primary fuels are used most in the UK for electricity generation? How can you tell from these bars?

d In what ways do you think these bars would have been different in 1950?

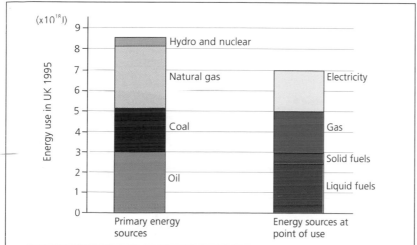

Figure 9.4

The pattern of primary energy sources and energy sources at the point of use, for the UK in 1995.

What are the trends in global energy use?

With the world population growing and people everywhere wanting as high a standard of living as they can achieve, the worldwide demand for energy is increasing quite rapidly. The main source of this energy continues to be fossil fuels. Figures 9.9 and 9.10 show the amounts of energy supplied, world-wide, between 1971 and 1996, by different fossil fuels and by other energy sources.

Figure 9.9

World Total Primary Energy Supply (TPES) by energy source.

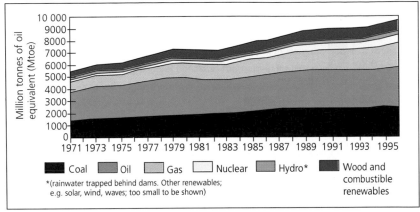

Figure 9.10

Where the world's energy came from, 1996.

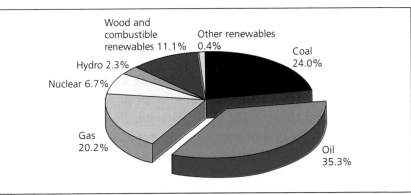

When looking at Figure 9.9, note that the amount of each energy source that was used in a particular year is shown by the vertical width of its band for that year. For example, in 1971, the amounts of each primary energy source used were as shown in Figure 9.11.

Figure 9.11 Quantities of each primary energy source, 1971	
Primary energy source	**Amount used in 1971 (in million tonnes of oil equivalent (Mtoe))**
coal	1400
oil	3800–1400 = 2400
gas	4700–3800 = 900
nuclear fuel	Very little
hydro	4800–4700 = 100
wood	5600–4800 = 800

Current and future demand for primary energy sources

We have already seen that the demand for primary energy sources varies considerably between richer and poorer countries. So if we are interested in predictions of demand for energy sources in the future, it is helpful to consider groups of countries, which are grouped geographically, or in terms of the extent of their economic and industrial development. The richest and most developed countries in the world, for example, belong to the Organisation for Economic Co-operation and Development (OECD), which includes most European countries (except those East European countries which were formerly part of the Soviet bloc), the USA and Canada, Australia, New Zealand and Japan.

Figure 9.12 shows the total amounts of primary energy used in seven regions in 1995 and the amounts that are predicted for 2020. The figures come from the International Energy Agency, which is an independent body linked to the OECD.

We will look later in the chapter at the kinds of assumptions that these predictions are based on. But for the moment, let's just consider their implications. We know, from Figure 9.9, that most of the world's energy comes from fossil fuels, or biomass (particularly wood). A little comes from hydro-schemes and nuclear fuels. Other renewable sources make a tiny contribution. Two obvious problems if energy use increases are:

- the finite supplies of fossil fuels might begin to run out; and

- the gases released when fossil fuels are burned already have a serious effect on the environment, and increasing these will make the situation worse.

The second problem is considered in more detail in Chapters 11 and 12. So here we will look at the first. ⊖ *Ref to Chapters 11 and 12*

Energy reserves and resources

All commentators agree that the main source of energy for the foreseeable future will continue to be fossil fuels. As long ago as the late 1960s, there was great concern about the remaining amounts of each type of non-renewable fossil fuel. It was feared that oil and natural gas would run out quite early in the 21st century. The Club of Rome – an international group of scientists, economists, industrialists and civil servants who started to meet in Rome in 1968 – published a report called *Limits to Growth* in 1972, which gave estimates of fossil fuel reserves (Figure 9.13).

Their predictions assumed that economic growth would continue at an average rate of 4– 5% per annum.

Even their more optimistic estimate of fossil fuel reserves – based on assuming that more would be discovered which would increase known

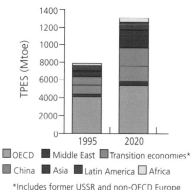

Figure 9.12

World TPES in 1995 and 2020 (predicted).

Legend:
☐ OECD   ■ Middle East   ■ Transition economies*
■ China   ■ Asia   ■ Latin America   ☐ Africa

*Includes former USSR and non-OECD Europe

Question

8

a By how many times is the energy consumption of each region expected to rise between 1995 and 2020?

b List the regions in order starting with the region in which there is the greatest proportional rise in energy consumption.

c Suggest reasons for the regions which come at the top and the bottom of your list.

Figure 9.13 Estimates of fossil fuel reserves, 1972.

Key term

The **reserves** of a fossil fuel are the total amount of that fuel which we believe to exist.

Figure 9.13 Estimates of fossil fuel reserves, 1972.

Fossil fuel	Years supply remaining based on known reserves	Years supply remaining based on known reserves x 5
coal	111	150
gas	22	49
petroleum	20	50

reserves five times – has turned out to be unduly pessimistic. Very considerable further reserves of fossil fuels were discovered during the final quarter of the 20th century – in the North Sea, in Alaska, and in already known oilfields. Another factor has been the rapid improvement of oil extraction technology, allowing companies to extract oil from reserves that could not have been used in the past. The current view of energy planners is clearly stated in a 1997 report to the British government (Figure 9.14).

Figure 9.14

From British Government Panel on Sustainable Development: Third Report, January 1997, paragraph 27.

27. The availability of primary energy, as distinct from its use, is unlikely to be a major problem for the first half of the next century: at present rates of production and usage, proven resources of fossil fuels are thought to be around 45 years for oil, 65 years for natural gas and 235 years for coal.

Questions

9 If fossil fuel reserves were five times as extensive, they would not last five times as long (see Figure 9.13). Explain why.

10 For each fossil fuel, compare the 'optimistic' Club of Rome estimates of reserves (Figure 9.13) with the estimates made in 1997 (Figure 9.14). (Remember that the Club of Rome estimates were made 25 years earlier.)

So while it is obvious that a finite resource such as fossil fuels must run out at some point – and their cost is likely to increase as they become scarcer – the general view is that the problems caused by burning fossil fuels are a more urgent problem than their supply.

Reducing the use of fossil fuels

Burning fossil fuels causes environmental problems on a global scale. As the highest per capita use is in wealthy countries, such as the UK, it is these that will have to make the biggest reductions in fossil fuel use, if total world consumption is to be significantly reduced.

Ref to Chapter 12

Figures 9.15 and 9.16 show the relative amounts of primary fuels used in the UK, and what these are used for. If we want to reduce consumption, then we need to look at some of the large sectors of use. Savings here will have the greatest impact. Electricity generation, and the contribution that renewable energy sources are expected to make towards this, are considered in Chapter 10. So here we will look at two other main areas of use: in homes and for transport.

Energy efficiency in homes

Science and technology is helping to reduce energy consumption in homes, for example by developing more efficient types of lamps, more efficient fridges and freezers and more efficient methods of cooking such as microwave ovens. But improved technology is not the only, or

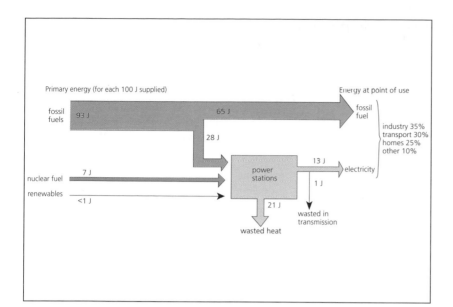

Figure 9.15

Energy use in the UK.

Primary energy (for each 100 J supplied)

fossil fuels 93 J

65 J

28 J

nuclear fuel 7 J

renewables <1 J

power stations

13 J electricity

1 J

21 J wasted heat

wasted in transmission

Energy at point of use

fossil fuel

industry 35%
transport 30%
homes 25%
other 10%

even the most effective, way of reducing energy consumption in homes. Energy consumption could also be reduced by more careful use of lights and of hot water. People could, for example, have showers rather than baths and boil no more water than is actually required when making hot drinks. Still larger amounts of energy could be saved by more effectively retaining heat in homes by means of effective insulation and draught-proofing. Indeed, if homes in the UK retained heat as well as many homes in Sweden, the waste heat from appliances would cease to be waste and would make a very significant contribution to keeping homes warm.

Question

11 From the information in Figure 9.15 and earlier in this chapter:

a What is the percentage of fossil fuels:

i in the total UK primary energy supply?

II In the energy sources used to generate electricity?

iii in the energy supplied at point of use?

b What is the average efficiency of power stations?

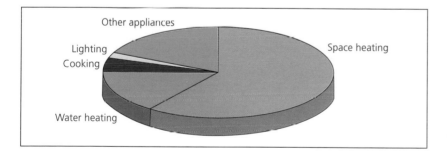

Other appliances

Lighting
Cooking

Space heating

Water heating

Figure 9.16

How energy is used in homes in the UK.

Discussion points

Could you reduce the amount of energy you use for heating in your home? Could you do this without making much difference to your level of comfort? What would be the most effective energy savings you could make?

One way of reducing energy use in homes is to increase VAT on fuels.

When this was tried a few years ago, the government was forced by public opinion to reduce the VAT level again. Should fuels be taxed more heavily, in order to reduce their use? If this were done, could ways be found to protect vulnerable groups? Are you in favour of higher taxes on fuels, or would you oppose this?

Figure 9.17

Passenger transport in Europe.

Fuel-efficient transport

Technology has made some contribution towards reducing the amount of fuel used for transport. More efficient engines give more kilometres per litre, and also reduce the amount of polluting gases that are released into the atmosphere. But, as with homes, really significant reductions in the amount of fossil fuels burned for transport cannot be achieved by science and technology alone. They require people to change their habits, both in the amount of travelling they do and the way they do it. Nowadays, the average distance travelled each day by a person in the UK is 28 miles (45 km) compared to just 5 miles (8 km) in 1950. Fig. 9.17 shows the changes between 1971 and 1997 in how people travel in the European Union. (Travel on foot and by cycle are too small to show on the graph but both have been decreasing over this period.) *Ref to Chapter 11*

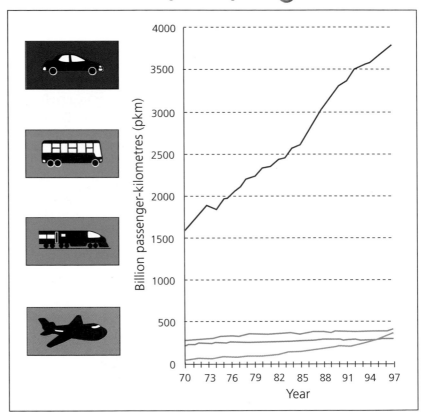

The most obvious way to reduce the amount of fossil fuel used for transport is for people to reduce the amount of travelling that they do. But some travel is necessary, and so people need to be encouraged to travel by the most fuel-efficient means. How can we compare the fuel-efficiency of different modes of transport? One way is to look at the number of kilometres a vehicle can travel per litre (or per gallon) of fuel. But it is more useful to compare them in terms of the number of passenger-kilometres, or tonne-kilometres, per litre of fuel (Figure 9.18).

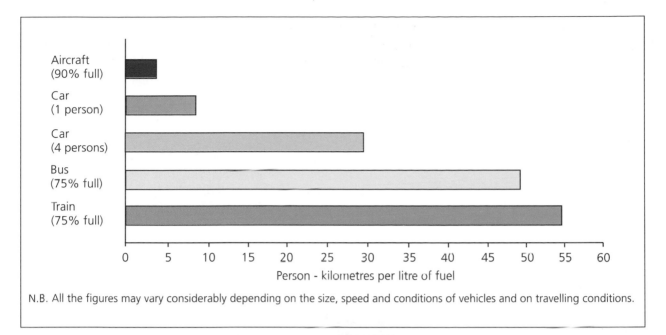

N.B. All the figures may vary considerably depending on the size, speed and conditions of vehicles and on travelling conditions.

Figure 9.18

Fuel efficiency of different modes of transport.

False economy

Modern cars have more efficient engines than older cars. So some countries, including Japan, have tax incentives to encourage the scrapping of older 'gas guzzlers'. However, manufacturing a new car means that you have to produce the materials from which the car is made, and this consumes a very large amount of energy. Researchers in the Netherlands have calculated that scrapping cars 3 years earlier than normal results in 4% *more* energy being consumed overall.

Discussion points

On balance, do you think the motor car has enhanced or reduced the quality of people's lives in the UK? List points on both sides of the argument before reaching your conclusion.

Can you think of any examples where an increase in standard of living might result in a reduction in quality of life?

Make a list of factors that you think would increase your quality of life. Make a similar list of factors that you think would reduce it.

Questions

13

a Which are the two most fuel-efficient ways of travelling (other than on foot or by cycle)?

b Which is the least fuel-efficient way of travelling?

c Under what circumstances is a car a reasonably fuel-efficient way of travelling?

14 The number of kilometres a car travels on a litre of fuel is reduced when the car is full of passengers, but not by very much. Suggest a reason why.

Key terms

Quality of life is not necessarily the same as **standard of living**. It isn't possible to measure quality of life in the way that standard of living can be measured using per capita GDP or energy consumption. It is possible, though, to identify factors which increase or reduce people's quality of life, though some of these are matters of personal choice about which individuals disagree.

Changing people's travel patterns

Despite the fact that public transport is more fuel-efficient than using a car, many people prefer to use their cars. There has been a huge increase in car travel during the final quarter of the 20th century. This is predicted to continue at a similar rate during the first quarter of the 20th century. It has important implications not only for energy consumption but also for the quality of people's lives. More cars on the road increases congestion and results in slower and more frustrating journeys. Many people spend several hours each day travelling relatively short distances to and from work. The pollution caused by vehicles also reduces the quality of life for many city dwellers much of the time.

Different points of view on transport matters are often strongly expressed in the media – on TV, on radio and in the press – in an attempt to influence public opinion and government policy. The statements below indicate some quite common points of view:

Figure 9.19

Contrasting opinions about public and private transport.

A People's quality of life is greatly enhanced by having the freedom, through owing a car, to travel where they want when they want. It is not the job of governments to restrict this freedom but to do whatever is needed to facilitate it.

B The evidence of the past 50 years indicates that building new roads does not ease traffic congestion but simply generates more traffic.

C In times gone by, before the mass ownership of the motor car, public transport was an efficient and much used way of getting about. The car has since undermined these more sensible methods of travel, but in doing so it may have become a victim of its own popularity. In many cities the average journey speed of the car driver is going down, not up. Car drivers spend more and more time in stationary queues, enjoying the so-called convenience of car travel surrounded by thousands of their kind each of whom is pumping exhaust fumes into the air from a uselessly idling engine. It is all a far cry from the kind of driving which is shown in car advertisements.

D In rural areas a private car is an essential part of life. Everyday things like getting to work, taking children to school and buying groceries would be quite impossible without one.

E Public transport is usually infrequent, unreliable and expensive making journeys a nightmare, especially for the elderly and those with small children.

Question

15 Match the statements A–E to the following people:

- a spokesperson for the Department of Transport and the Environment;
- the authors of the book *Blueprint for a Green Planet*;
- a parent who lives in the country;
- a parent who lives in a city;
- a spokesperson for a motoring organisation.

Predicting future energy use

Figure 9.12 presents one prediction about future energy demand. As one energy expert, however, has commented: 'Energy forecasting is easy. It's getting it right that's difficult.' All forecasts are based on assumptions. The commonest assumption that people make is:

• The future will be like the past.

In other words, trends up to the present are likely to continue. On the other hand, we are not powerless to change things. We can also say that:

• The future is what we (collectively) choose to make it.

Scientists working on energy policy base predictions of energy demand on *scenarios*. Originally the word 'scenario' meant the outline of a play or film. Here it means a set of assumptions that build up a picture of what the future might be like – which can be used to construct a model (usually a mathematical model, that can form the basis of a computer program) for predicting future trends.

One obvious scenario is to assume that the first point above is our best guide. On that basis we might predict a growth of around 2.5% in energy demand. This is higher than the growth of population, so it allows for some catching up by the poorer countries. Oil consumption would rise till 2020 and then fall as stocks became lower. Gas would rise and then level off; coal, too, would rise. More hydroelectricity and nuclear power, and greater use of biomass fuels, would be needed to satisfy demand. From 2020, wind and solar energy would increase rapidly, until they contributed more than the total energy consumption today. This is summarised in Figure 9.20.

A rather different scenario assumes that worries about global warming, along with dwindling fossil fuel stocks, will lead the world to decide to phase out fossil fuels entirely within 100 years; and concerns about nuclear power lead to its too being phased out. All energy requirements then would come from biomass (about 50%) and other renewable sources (the remaining 50%). This would involve capping consumption in the developed countries at the present level, while allowing consumption in the poorer countries to rise faster than their populations. Gas (which releases less CO_2 per unit of energy) would be used to tide us over until the contribution of renewables had been increased – and then gradually reduced. This is summarised in Figure 9.21.

Ref to Chapter 12

The reality may turn out to be somewhere between these, and other scenarios are being used to model fuel consumption, using other 'compromise' assumptions. The main thing to appreciate is

Figure 9.20

Predicted world primary energy consumption to 2050: based on 'no change' scenario.

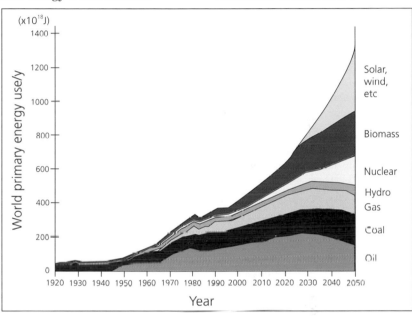

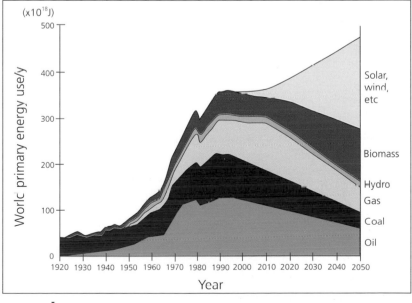

Figure 9.21

Predicted world primary energy consumption to 2050: based on 'green' scenario.

that all future predictions:

- are subject to considerable uncertainty;
- are only as good as the assumptions they start from, and those built into the model.

And also to recognise that the second bullet point at the start of this section applies. The future *is* up to us.

Discussion points

What do *you* think is the most likely scenario for energy consumption until 2050? How would it differ from the two scenarios outlined on page 127?

From the point of view of global energy supply, do you think the following government policies are sensible:

a Reducing VAT on domestic fuels?

b Introducing a lower annual road tax for cars with smaller, more economical engines?

If you were an adviser to the Minister of Transport, what would you advise her (or him) to do to reduce fuel use for transport in the UK? Would you use incentives to encourage people to use public transport or discourage car use by raising its cost? How would you go about persuading people to change their transport patterns? Your answer should take account of different points of view and the differences between living in a city and in the country, and the special requirements of the elderly and those with physical handicaps. Your answer might refer to some of the following ideas: bus lanes; bicycle lanes; keeping cars out of city centres or charging them to go there; more efficient and cleaner engines; a major programme of road building; greatly increasing the tax on fuel; subsidising frequent and reliable public transport. You should also indicate the order in which your suggested measures are best implemented.

Review Questions

16 Use Figure 9.22 to answer the following questions.

a Which region has the highest per capita TPES?

b Which two regions have a per capita TPES less than half of the world average?

c Which region has the lowest TPES:GDP ratio?

d Which region has the highest TPES:GDP ratio?

e If there were a perfect correlation between energy consumption and standard of living TPES:GDP ratio in different regions of the world would be exactly the same. Use the idea of efficiency to explain why they are not.

f About a fifth of the world's population live in China. The increase in population is already being tightly controlled. What else needs to be done in China if improving the standard of living in that country is not to cause a huge increase in the consumption of fossil fuels?

17 Trains, buses and aircraft often last for 20–30 years compared to an average of about 10 years for a car. They also travel many more miles each year than an average car. Explain how this helps to make them more energy-efficient.

Figure 9.22 Total primary energy supply (TPES) per person and the ratio between TPES and GDP in different regions of the world.

Region (1996)	TPES/Pop (toe/capita)	TPES/GDP (toe/000 US$)
World	1.68	0.39
OECD	4.60	0.26
Middle East	2.04	0.62
Former USSR	3.19	1.91
Non-OECD Europe	2.01	0.88
China	0.90	1.57
Asia	0.57	0.68
Latin America	1.09	0.39
Africa	0.62	0.89

Chapter 10

Electricity supplies

The issues

At present, most of the world's electricity is generated using the energy released from burning fuels, but this is having a serious effect on the Earth's atmosphere. In some high-income countries, a significant proportion of electricity is generated in nuclear power stations. Although these do not release large quantities of polluting gases into the atmosphere, they cause other environmental problems, because of the radioactive waste they produce, so there is an increasing interest in the use of renewable energy sources to generate electricity. Until recently, however, electricity from renewable energy sources has been much more expensive. Furthermore, methods of generating electricity from renewable energy sources also have environmental impact.  *Ref to Chapter 12*

If people want all the benefits of a readily available supply of electricity, they must balance the cost of each method of generation against its environmental impact, to decide which methods to use (Figure 10.1).

The science behind the issues

Electricity is a secondary energy source: it has to be generated using a primary source of energy e.g. a fossil fuel, nuclear fuel or a renewable energy source such as solar radiation, wind or waves. This usually involves using the primary energy source to turn a turbine which then drives a generator. Solar cells, however, produce electricity directly from sunlight.

With renewable energy sources there are no fuel costs. However, they are less concentrated sources of energy than fuels. As a result, the capital cost of the equipment needed to harness their energy is much greater, per unit of electricity generated, than for fuel-burning power stations.

Figure 10.1

The way we live our lives depends to a large extent on a readily available supply of electricity.

What this tells us about science and society

Decisions about how to generate electricity involve weighing up technical, environmental and economic considerations. Technological improvements to a method of generation can reduce its cost or its environmental impact. Governments may wish to see renewable energy sources used more extensively, to honour international commitments. Government policy must also take account of public opinion about matters such as the health risks of nuclear power stations or where power stations of any type should be located. Science can inform these decisions, but cannot determine what they should be.

1 Imagine that there is a 24-hour power cut. Make a list of the things that you normally do each day that you wouldn't then be able to do.

2 In what way(s) is the sequence of steps involved in generating electricity in a nuclear power station the same as that in a coal-fired power station? In what way(s) is it different?

3 In what way(s) is the sequence of steps involved in generating electricity in a gas-fired power station different from that in a coal-fired power station?

4 In what way is generating electricity from solar cells different from generating electricity from all the other energy sources shown on Figure 10.2?

5 For each of the primary energy sources shown on Figure 10.3:

a describe the trend in the amount of electricity generated during the period 1990–98;

b suggest a reason for the trend.

6 What percentage of UK electricity was generated in 1998 using fossil fuels as the primary energy source?

How are primary energy sources used to generate electricity?

Figure 10.2

Figure 10.2 shows the steps involved in generating electricity from the various primary energy sources that are most often used.

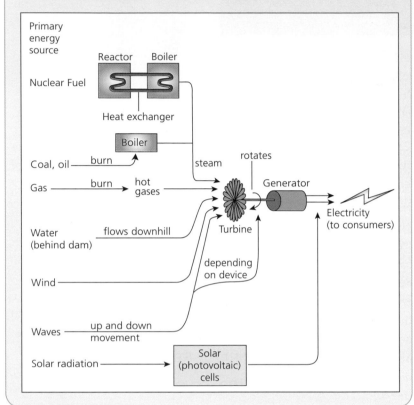

What energy sources are used to generate electricity in the UK?

Figure 10.3 shows how the amount of electricity generated in the UK from each of the four main sources has changed since 1990.

Key term

The unit normally used to measure amounts of energy supplied by electricity in domestic situations is the **kilowatt-hour** (kWh). It is the amount of energy required to run a 1 kW device for 1 hour.

1 terawatt-hour (TWh) = 1 billion (10^9) kWh.

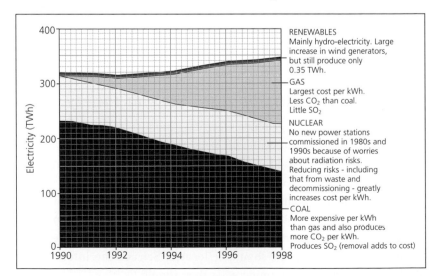

Figure 10.3

Energy sources used to generate UK electricity.

RENEWABLES
Mainly hydro-electricity. Large increase in wind generators, but still produce only 0.35 TWh.

GAS
Largest cost per kWh. Less CO_2 than coal. Little SO_2

NUCLEAR
No new power stations commissioned in 1980s and 1990s because of worries about radiation risks. Reducing risks - including that from waste and decommissioning - greatly increases cost per kWh.

COAL
More expensive per kWh than gas and also produces more CO_2 per kWh. Produces SO_2 (removal adds to cost)

Power stations that use fuels as their primary energy source are not very efficient. Gas-fired power stations can achieve an efficiency of about 50% but coal-fired power stations and nuclear power stations, both of which use steam to drive turbines, are at best 40% efficient. Since most electricity in the UK is generated using fuels, it is important to consider how, if at all, fuel-burning power stations could be made more efficient. Ref to page 118

How can fuel-burning power stations be made more efficient?

In a fuel-burning power station, heat is produced when the fuel is burnt and this is used to boil water to make steam. This then drives a turbine. Once the steam has passed through the turbine blades, it is condensed back to water, to draw more steam through. The cooling water used to do this is heated in the process (Figure 10.4). In any fuel-burning power station, a lot of the heat produced by burning the fuel is carried away by this cooling water. As a result, the efficiency of the power station is low (Figure 10.5).

The efficiency of fuel-burning power stations can be increased to some extent by improved technology, for example, with better turbine design. But no device that uses heat to produce movement (no 'heat engine') can ever achieve anything approaching 100% efficiency – for theoretical reasons. The reason is that some of the heat always has to be transferred to a cold reservoir (like the cooling water) to keep the process running. And this energy is wasted. In fact, the maximum theoretical efficiency for a technologically perfect heat engine using steam at 500°C is around 60%.

Although the cooling water from a power station is at too low a temperature to be of any use for generating electricity, it *can* be used for heating buildings. Schemes which use the waste energy in this way are known as combined heat and power (CHP) (Figure 10.6). It is difficult to do this with large power stations because they are usually sited well away from large centres of population. But an increasing number of smaller

Figure 10.4

Steam rising from cooling towers at the Didcot power station in Oxfordshire. Cooling towers have been described as monuments to the inefficiency of large power stations.

Question

7 Draw a diagram similar to Figure 10.5 for a gas-fired power station.

Figure 10.5

Efficiency of an ordinary power station.

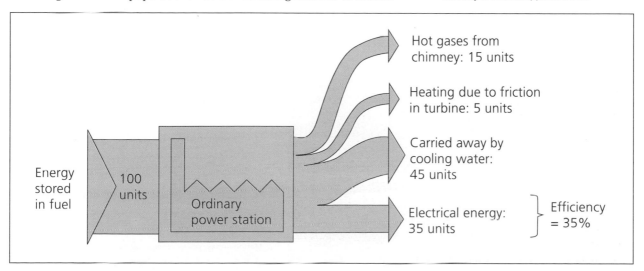

Energy stored in fuel — 100 units — Ordinary power station

Hot gases from chimney: 15 units

Heating due to friction in turbine: 5 units

Carried away by cooling water: 45 units

Electrical energy: 35 units

Efficiency = 35%

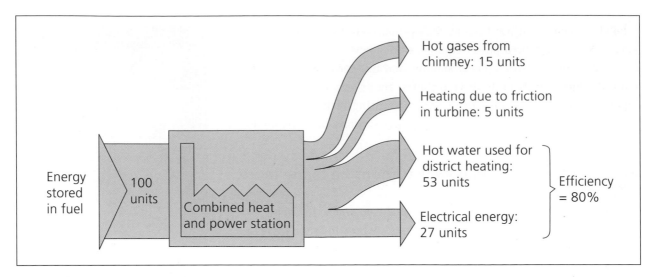

Figure 10.6

Efficiency of a CHP scheme.

CHP schemes are being built, which generate electricity for a local community or for a single hospital, or factory, or university – and use the cooling water to provide central heating for the buildings around.

Are nuclear fuels greener than fossil fuels?

Most electricity in the UK is generated using fuels, and this is likely to remain the case in the foreseeable future. Since burning fossil fuels inevitably releases large quantities of carbon dioxide into the atmosphere, nuclear fuels offer what appears to be a 'greener', more environmentally friendly option (Figure 10.7). This advantage of nuclear power has been emphasised by the nuclear industry. They point out, for example, that existing nuclear power stations reduce the amount of carbon dioxide being released into the atmosphere by as much as it would be if 70% of cars were taken off UK roads. They also draw attention to the fact that France, over a 10-year period during the 1980s and 1990s, increased the nuclear share of electricity from 20% to 70% and so reduced carbon dioxide emissions by about 60%. In the UK, no new nuclear power stations were commissioned during this period, though some already under construction were completed.

The main reason for the different policies on nuclear power in the UK and in France is the far greater public concern in the UK – rightly or wrongly – about the health risks of nuclear power. Nuclear power stations use

Figure 10.7

Making nuclear power seem a 'natural' option.

highly radioactive materials and the radiation that these emit can seriously damage people's health. ⟳ *Ref to chapter 13*

In addition to any risks to workers in the nuclear industry itself, there are three possible sources of radiation risk to the general public:

- from the routine leakage, however slight, of radioactive materials into the immediate environment. This mainly affects people who live close to nuclear power stations or to plants which process nuclear fuel.

- from the waste produced by nuclear power stations, some of which has a long half-life and will remain dangerously radioactive for hundreds of years. The people who might be at risk from this depends on where, and how, the waste is stored.

- from accidents which release radioactive materials into the environment. Depending on the scale of the accident and weather conditions the effects of this might be local or spread over a wide geographical area.

These sources of radiation risk will be considered in turn.

Low-level leakage from nuclear plants

The main concerns about any routine leakage of radioactive materials into the immediate environment of a nuclear power plant are about the levels that are permissible and the extent to which the radioactive materials accumulate. Regulations about permitted levels have been progressively tightened up since the introduction of nuclear power stations in the late 1950s and only materials with a relatively short half-life may be released. Nevertheless, these may become more concentrated as they move up the food chain.

The problem of radioactive waste

Nuclear power stations produce three main types of radioactive waste:

- high-level   spent fuel rods; solidified waste from reprocessing fuel rods;

- intermediate-level   cladding that surrounds fuel rods, filters and residues from effluent treatments; worn-out items of equipment from nuclear power plants;

- low-level   protective clothing; rubble and steelwork from decommissioned nuclear power plants.

At present most of the high-level and intermediate-level waste – which is hot both in temperature *and* in the sense that it emits dangerous radiation – is stored in constantly cooled tanks at the reactor site where it was produced. This type of storage is only designed to be short term, and some more permanent storage will eventually be needed.

Figure 10.8 ▌

Britain's main nuclear energy site is located on a remote part of the Cumbrian coast.

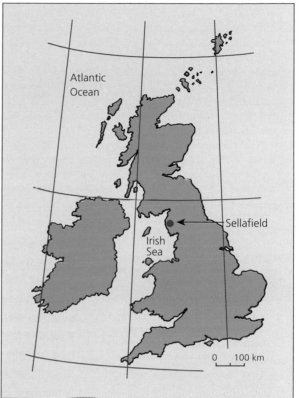

Atlantic Ocean

Sellafield

Irish Sea

0   100 km

Discussion point

Explain what it means to say that a radioactive substance 'may become more concentrated as it moves up the food chain'. How might this result in fish caught in the Irish Sea causing a significant health risk?

Figure 10.9

Dealing with nuclear waste
(from New Scientist, 5 October 1996).

The following article (Figure 10.9) explains the ideas that have been suggested for the safe storage of nuclear waste in deep underground repositories.

First, the waste is sealed in a glass matrix which is highly resistant to radiation damage, binds radionuclides such as uranium and plutonium, and corrodes very slowly. The glass matrix is sealed into steel cylinders which should isolate the waste for a minimum of 1000 years. the canisters are designed so that their corrosion products include iron oxyhydroxides which act as a chemical buffer that absorbs radionuclides.

The canister is surrounded by a backfill of clay that slows diffusion of dissolved radionuclides. The final barrier is geological: the repository will be built deep in rock that will contain any escaping radionuclides, that has little movement of groundwater, and that can offer physical protection to the engineering barriers.

In deep repositories for low and intermediate level waste, radionuclides would be physically restrained in steel drums filled with a highly alkaline grout–a special type of cement. The high pH is important because radionuclides tend to be less soluble in alkaline conditions than in neutral or acidic media. The backfill used to surround the steel drums will also be alkaline. Once again, the site of these repositories should be chosen so that the radionuclides take tens of thousands of years to reach drinking water aquifers, by which time the levels of radioactivity should have decayed to safe levels.

Question

9 Draw a labelled and annotated diagram to show the proposed method for the safe storage of high-level radioactive waste.

Discussion points

Sellafield is the largest nuclear power plant in the UK and is by far the biggest employer in its area.

a Describe the region in which it is located (see Figure 10.8).

b Some local people supported the proposed exploratory repository near Sellafield. Suggest why.

How effective do you think the type of protest shown in Figure 10.10 would be in influencing:

a the person conducting the public inquiry?

b members of the general public?

Give reasons for your answers.

So far, however, no country has managed to find a safe underground site for its high-level or intermediate-level nuclear waste. In the UK, for example, the Nuclear Industry Radioactive Waste Executive (Nirex) made a planning application in 1990 for an exploratory underground repository 300 metres underground at Longlands Farm, near to the British Nuclear Fuels Limited (BNFL) plant at Sellafield. At the public enquiry, held in 1996, it was not only a group of local residents who opposed the Nirex plans (Figure 10.10). Some geologists argued that Nirex had inadequate knowledge of the complex geology of the proposed site and drew attention to the fact the 450 million years-old rocks were riddled with hairline fractures, often hundreds of metres long. This meant, they maintained, that it was impossible to predict the flow of groundwater which might well become contaminated with radioactive substances from the waste. Nirex responded to these objections by claiming that the

Figure 10.10

Protesters pulling a Trojan horse through Cleator Moor in Cumbria on the first day of the inquiry into the planning application to build an underground repository for nuclear waste.

best way to resolve the geological issues, and test the safety of the site for a full-scale repository, was to build the test repository. In March 1997, Nirex's planning application was rejected on the grounds of scientific uncertainties and technical deficiencies.

Nuclear accidents

A major reason for opposition to nuclear power is the fear of serious accidents and the scale of radioactive contamination that these might cause. This fear is based on two sorts of reason.

First, there have, in fact, been quite a number of nuclear accidents. The two most serious of these, in terms of the amount of radioactive pollution released, were the fire and resulting leakage at Windscale (now known as Sellafield) in England in 1957, and the explosion at Chernobyl in the Ukraine in 1986.

Secondly, with any complex technological process, it is never possible to anticipate all of the things that might conceivably go wrong or to make every aspect of the process completely fail-safe. Nor can we ever be completely sure that any software that is used to control the process is completely free of 'bugs'.

Furthermore, *people* are involved at various stages in the process and it is never possible to rule out the possibility of human error. Nor is it desirable to eliminate human involvement in the technological process altogether because, if something happens that was entirely unforeseen by the designers of the plant, a sensible response to the situation will necessarily involve human judgement. Unfortunately, however, neither the judgement that something has happened that requires human intervention, nor the judgement about how precisely to intervene, can be guaranteed to be correct. In science and technology, as in life generally, there can be no absolute certainties. The possibility of accidents in nuclear plants can never be entirely eliminated. In fact, human error played a significant part in both the Windscale and the Chernobyl accidents.

Figure 10.11

Radiation doses in OECD countries resulting from the Chernobyl accident (compared to the average annual dose from natural sources).

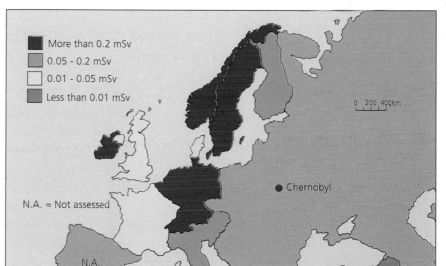

- More than 0.2 mSv
- 0.05 – 0.2 mSv
- 0.01 – 0.05 mSv
- Less than 0.01 mSv

N.A. = Not assessed

0 200 400km

● Chernobyl

N.A.

10 In October 1999, an accident occurred at a nuclear plant at Tokaimura in Japan. The accident was of a well-known type that has occasionally occurred elsewhere. If there is enough enriched nuclear fuel together in the same place – more than what is called a critical mass – an uncontrollable chain reaction occurs, causing an explosion. This type of accident is called a criticality accident.

The following extract is taken from the leader in the science magazine *New Scientist*, the week after the accident occurred:

'The runaway chain reaction at Tokiamura last week cannot be blamed on fractured pipes, equipment failures or software glitches. Nor were workers fighting for their lives. ...They were just following a routine ... that positively invited a criticality accident.

'Simply to blame the workers makes no sense. If the managers and regulators of nuclear plants can't make them immune to obvious sources of human error ... they are not doing their job. ... Nobody else now uses a wet process to make fuel rods from highly enriched uranium ... they use a dry process. ... The reason is quite simple: water facilitates criticality.'

New Scientist, 9 October 1999

a Give one reason why the routines at Tokaimura 'positively invited a criticality accident'.

b Who does the writer of the article mainly blame for the accident and why?

Figure 10.12

Pipes leading to a hydroelectric power station at the aluminium works near Fort William, Scotland. The pipes carry water from two lochs about 20 km away. Most of the UK sites that are suitable for large-scale hydroelectric schemes are already being used.

How will the UK's electricity be generated in the 21st century?

Since no new nuclear power stations are planned, and many existing nuclear power stations are reaching the end of their useful lives, the amount of electricity generated in the UK from nuclear fuels will decline during the early years of the 21st century. The only alternative to fossil fuels will come from renewable energy sources.

The UK government is, in fact, committed by an international agreement to increasing the amount of electricity generated from renewable energy sources to 10% by 2010. The figure in 1999 was 2.5%, mainly from hydroelectric schemes. There is not much scope for increasing the contribution from hydroelectricity, as virtually all the suitable UK sites for large-scale hydro-electric schemes are already in use (Figure 10.12). A similar way of generating electricity using water involves building a barrage across a river estuary. Sea water is trapped behind the barrage at high tide and flows out through turbines as the tide falls. Building a barrage across the Severn estuary could provide 7% of UK electricity and so go a long way towards meeting the government's commitment to renewable energy sources. The barrage would, however, destroy a habitat that is important for wading birds and is, therefore, strongly opposed by organisations such as the RSPB (Royal Society for the Protection of Birds).

One reason why the move towards other renewable energy sources has been very small and very slow is that the cost of each unit (kilowatt-hour, kWh) of electricity generated using these energy sources has tended to be greater than for conventional power stations (Figure 10.13). In 1989, to encourage the development of renewable energy technologies, the government set up the Non-Fossil Fuel Obligation (NFFO). This requires electricity companies to buy a small proportion of their electricity from non-fossil fuel sources. The money for this comes from a levy which is charged to every electricity bill and which raises almost £1 million each year. To begin with, much of this money went to the nuclear industry, but from 1997 a greater share has gone to renewable energy projects.

This example, of course, illustrates an important general point: the cost of generating a unit of electricity by each of the available methods is not determined solely by technical considerations, such as the efficiency of the generating equipment. Government policy on taxation of fuels, and subsidies or tax breaks for certain fuels, or for the disposal of wastes, can make some methods appear cheaper than their true cost. But this could change in the future, if policies changed, or if political or economic factors, perhaps beyond the UK, led to the costs of some fuels rising or falling dramatically.

Initiatives for a greater and faster move towards renewable energy sources have not only come from government. It became possible, in 1998, for customers to buy electricity from any supplier, rather than from just one regional electricity company. A new electricity supply company, *unit(e)* was set up to supply electricity generated exclusively from renewable energy sources. This company already has contracts with several wind farms and hydroelectricity schemes. It plans to increase its portfolio of suppliers as customer numbers grow and thus stimulate the development of new projects for generating electricity from renewable energy sources. The electricity supplied by this company costs slightly more than that supplied by the major electricity supply companies (about £3 per month on a typical household bill).

If the real cost to society of the waste materials produced by fuel-burning methods were added into the electricity bill, it would alter the relative costs of different generating methods, and might well change patterns of generation and consumption. Until this policy change occurs, cost is likely to remain a significant influence on people's choices. So those in favour of renewable energy sources are working hard to improve the technology so that the cost of each unit of electricity generated is more or less the same as the electricity generated by power stations which use fuels as their primary energy source – even under the current economic rules.

Figure 10.13 Cost of generating electricity using renewable energy sources, 1995.

	Cost (pence/kWh) UK, 1995
Gas	~ 2.5
Waves	2.6 – 6.0
Wind	3.0 – 4.5
Coal/oil	~ 4.0
Solar (titanium oxide, amorphous silicon)	~ 4.0
Nuclear	4.5+
Solar (crystalline silicon)	~ 20

Key: + plus (large) costs of decommissioning

Discussion point

In the UK, there is a National Grid for transmitting electricity. All the generators feed electricity into the grid, and customers draw electricity from the grid. In what sense, then, are *unit(e)'s* customers buying electricity that has been generated from renewable energy sources?

Question

11 Make a list of the main arguments for increasing the amount of electricity in the UK generated from renewable sources. Then list the main reasons why the share is still low.

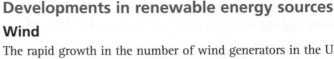

Discussion points

Assess the advantages and disadvantages of emulating the Danes in their push towards wind power.

Many people are in favour of wind farms provided that they are not too near to where they live. This attitude is called NIMBY (not in my back yard). How, and by whom, should the siting of wind farms be decided?

Developments in renewable energy sources

Wind

The rapid growth in the number of wind generators in the UK during the 1990s, from a mere handful to nearly 800, was partly due to the NFFO described on page 137. Considerable reductions in the cost of each unit of electricity generated were also achieved during the same period, however, by economies of scale. That is, once a particular design of wind generator has been successfully developed, and wind generators of that design are then manufactured in substantial numbers, the cost of each individual wind generator is considerably reduced.

The growth in the numbers of wind generators in the UK is likely to continue into the 21st century. There is, however, a limit to the proportion of UK electricity that can be generated in this way. It has been estimated, for example, that if every wind generator that existed in the world at the start of the 21st century were moved to the UK, they might just about produce 10% of UK electricity, but only on a windy day. In fact, as with hydroelectricity schemes, most of the suitable sites are likely to be used quite quickly. Furthermore, what counts as an appropriate site is itself a disputed matter. Wind generators are usually built in groups (wind farms) and are often located on the tops of hills where they can be seen for miles around. Some people feel very strongly that they are an eyesore and oppose them when planning permission is applied for (Figure 10.14).

Just as the French have gone very strongly for nuclear power, the Danes are concentrating on wind power and aim to generate half of their electricity from wind by 2030. They already have over 4000 medium-sized (500–750 kW) land-based turbines which have a considerable impact on the environment. The next phase of their programme, to 2008, is to build 500 larger (1.5 MW) turbines off-shore, in shallow coastal waters.

Waves

The harnessing of energy from waves for generating electricity (Figure 10.15) is not so well advanced as harnessing energy from the wind. The story of research into wave-energy devices in the UK during the final quarter of the 20th century illustrates the importance of funding for research, and how this funding may be affected by political considerations.

Britain's state-funded research programme into wave-power was cancelled, controversially, in 1982. The government claimed that waves were a wholly uneconomic source of energy. The scientists involved, however, accused ministers of sabotaging their research to protect

Figure 10.14

A wind farm at Penistone in South Yorkshire. Sculptural shapes or a blot on the landscape? 500-1000 wind generators are needed to replace one medium-sized fuel-burning power station. Even then the output is less reliable because the wind does not blow strongly all the time.

nuclear power, the real costs of which were then becoming apparent. Alternative funding had to be sought from private industry and from the European Commission.

Researchers continued to make considerable progress mainly using scale models in large wave-tanks. There was another major set-back however, when one of the more promising devices was tried out 'for real' (Figure 10.16).

TWO concrete blocks on the bottom of the sea off the north coast of Scotland are all that's left of the world's first attempt to build a commercial wave power station. When Osprey, a large yellow 2-megawatt generator, was wrecked by waves that were meant to power it, hope died. Before its steel ballast tanks could be filled, heavy seas scoured the sand from beneath them and they ripped open. The engineers who designed the machine were 'absolutely gutted' and Lloyds insurers had to pick up a bill for more than £1 million.

Osprey's spectacular failure in 1995 seemed to confirm almost everyone's prejudices about wave power. Government felt vindicated for having refused to fund it, on the grounds that marine engineering is just too costly, while conventional electricity producers, who have long ridiculed wave power, looked smug.

Despite the set-back, research continued. In 1997, Ian Thorpe, a consultant who had advised the government on wave-power since 1989, was able to inform the Department of Trade and Industry (DTI) that three types of wave-energy device could now generate electricity for under 6 p per unit. One of these devices, the nodding duck invented by Stephen Salter of the University of Edinburgh, could produce electricity for as little as 2.6 p/unit. This meant that electricity from wave-power was now genuinely competitive with that generated by conventional power stations.

So, after neglecting this source of renewable energy for 16 years, the British government was once again prepared to support it. Potentially, 20% of Britain's electricity could be generated from wave-power.

Solar cells

Solar cells have been used for many years to generate electricity directly from the Sun's rays. The cells are usually made from crystalline silicon which means that manufacturing them is not only quite expensive but also requires a considerable amount of energy. Furthermore, the cells are only 13–16% efficient (that is, only 13–16% of the energy in the radiation that falls on them ends up as electricity). Solar cells of this type are, therefore, only economic where they can be used either instead of batteries or to re-charge batteries, for example, on satellites, in remote situations on Earth or in watches and calculators.

Solar cells can also be made from amorphous silicon, that is silicon in which the atoms are not all neatly arranged as they are in the crystalline form. These cells are much cheaper to make but are considerably less

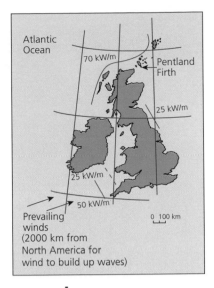

Figure 10.15 |

Potential for wave power in the UK.

Figure 10.16 |

The fate of a wave energy experiment. From New Scientist, *30 October 1998*

Question

12 With reference to Figure 10.15:

a explain why the UK is well situated to exploit wave energy;

b estimate the potential capacity (in TW) for electricity generated from wave energy around UK coasts.

(1000 kW = 1 MW; 1000 MW = 1 GW; 1000 GW = 1 TW)

Figure 10.17

Solar cells cover the upper surface of *Sky Blue Wabeda II*, a competitor in the 1996 World Solar Car Challenge.

efficient. They also gradually become degraded by light. During the 1990s, however, researchers have found ways of increasing the efficiency of these cells and limiting their degradation so that they are about 13% efficient even after 1000 hours of use. This means that the cost of the electricity they produce is similar to that of the electricity produced by burning fossil fuels.

Other researchers have developed solar cells from titanium oxide (a white solid that is often used in paint) and a dye. These cells are more efficient than silicon cells and cost only one-fifth as much. They can also be made from very thin transparent films which can be laid down on to glass, for example, the cover of a watch. The cell will then power the watch but the watch face is still visible through it. The windows of buildings could be covered with solar cells of this type.

Electricity generation in developing countries

Many of the people in low-income, developing countries live in small villages in remote areas (Figure 10.18). Generating electricity in large power stations and then distributing it to these villages, as happens in developed countries like the UK and in large cities in developing coun-

Figure 10.18(a)

Women pulling water from a well in a village in Senegal, Africa.

AN AFRICAN VILLAGE
The weather is generally hot, dry and sunny. Water has to be lifted from a deep well.

Figure 10.18(b)

A NEPALESE VILLAGE
The land is hilly with fast streams which have cut quite deeply into the ground

tries, is uneconomic. The cost of installing the power lines would be far too great, particularly in relation to the amount of electricity people could afford to consume.

Nevertheless, access to some electricity can significantly enhance the quality of life of the people who live in these rural areas. For example, having at least one community building that can be lit at night would allow many additional activities to take place and the availability of radio or TV would provide useful information, for example about the weather or other matters affecting agriculture, and provide education which would otherwise be unavailable.

What is needed in such situations is a method of generating relatively small amounts of electricity cheaply and reliably using technology that is appropriate to local circumstances.

Discussion point

Some methods of generating small amounts of electricity are shown below. Which would be most suitable for the African and Nepalese villages shown in Figure 10.18? Why is it best? What disadvantages might it have?

Petrol/diesel generator

Costs (250 W) £300

Needs regular, skilled servicing

Needs fuel which is expensive and has to be transported.

Array of solar cells

Costs (250 W) £1500

No moving parts, no servicing needed.

No fuel needed.

Only works in daylight and full output needs bright sunlight.

Micro-hydro

Cost (250 W) £500

Requires regular but simple servicing.

No fuel needed.

Requires reliable supply of running water and/or small dam.

a In 1998, just 6% of UK electricity was generated by CHP schemes. Why do you think this figure was so low?

b List the advantages and disadvantages of:

i CHP schemes involving medium-sized power stations located on the outskirts of towns and cities;

ii small, gas-fired CHP schemes located on housing estates or in the basements of blocks of flats, connected to the National Grid, and feeding in electricity or extracting it as necessary.

Review Questions

13 Use the graph in Figure 10.19 to answer the following questions:

a How do the renewable energy sources compare, as far as carbon dioxide production is concerned:

i with fossil fuels?

ii with each other?

b Why does electricity generated from renewable energy sources and nuclear fuels produce any carbon dioxide at all?

14 The following sentence is taken from the British Government Panel on Sustainable Development, Third Report, January 1997, Paragraph 27:

'The combined effects of environmental factors, technical innovation and financial pressures ....could lead to a much less carbon-intensive energy system, based on efficient, decentralised electricity generation relying heavily on renewable energy sources.'

This sentence is unlikely to mean very much to most members of the general public. Use what you have learnt in this chapter to write a few sentences which say the same thing as the above sentence in language that most members of the general public could easily understand.

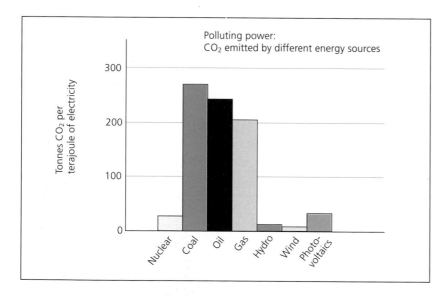

Figure 10.19

Amounts of carbon dioxide released into the atmosphere when electricity is generated using different energy sources.

Chapter 11

Air quality

The issues

People worry enough about air pollution for weather forecasters to include reports about the quality of the air we breathe. Air quality matters to people with heart conditions or lung disease such as asthma and bronchitis. Every hour of every day the Department of the Environment in the UK updates its information service aimed at anyone whose breathing gets worse when air pollution increases.

Not that worries about air quality are new. As long ago as 1273 the use of coal was banned in London because it was harmful to health. Today, however, the main sources of air pollution that affect air quality are motor vehicles and local factories.

The science behind the issues

There are many air pollutants but it is not obvious how they can all come from burning fuels. We have to know something about chemical reactions to understand why motor vehicle engines, power stations and factories are responsible for air pollution. When fuels burn at a high temperature the waste fumes include gases such as oxides of nitrogen, carbon monoxide, sulphur dioxide and sooty particles. These gases do not just disappear. They spread out into the air and can react with each other as well as with oxygen and water vapour in the air.

What this tells us about science and society

It is all too easy to accept the information about air quality without questioning the evidence on which it is based. When trying to interpret air-quality data we have to know where the scientists collect air samples for analysis and also have some sense of the accuracy and precision of the instruments used for taking measurements.

The issues of air quality shows that there is a tension between the rights of individuals and the wish of groups to regulate harmful activities. Many people want to own and drive motor vehicles but cars are one of the main sources of the air pollution which threatens everyone's health. So the freedom to drive has to be balanced against the harmful affects of motoring which affects us all. As a society we have to balance the benefits of individual freedom against the collective wish to breathe pure air. Governments have to decide on an appropriate and feasible strategy to improve air quality without stifling economic growth.

Figure 11.1

This cyclist in London is wearing a mask designed to filter out harmful gases and particles from the air he is breathing and seems to be more worried about air pollution than the risk of head injury.

Why does burning fuels cause pollution?

Burning fuels in engines

Burning fossil fuels is the main source of many pollutants. Only very few of the hundred or so chemical elements are involved. Petrol, for example, consists almost entirely of a mixture of hydrocarbons. These are compounds of just two elements: hydrogen and carbon. When petrol burns in an engine the main reactions involve hydrocarbons reacting with oxygen from the air. The hydrogen atoms in the fuel join with oxygen to form hydrogen oxide (water) while the carbon atoms join with oxygen to form carbon dioxide. This combustion reaction releases energy.

Figure 11.2

Chemists use models of molecules to describe what happens during chemical reactions. In this example a hydrocarbon reacts with oxygen forming water and carbon dioxide. The atoms of the elements do not change but they rearrange to make new molecules.

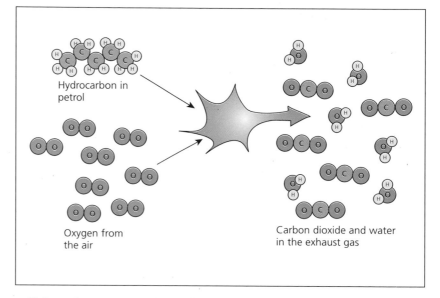

Hydrocarbon in petrol

Oxygen from the air

Carbon dioxide and water in the exhaust gas

If the only gases in exhaust fumes were carbon dioxide and water we would not have to worry about their effect on air quality. Carbon dioxide is a greenhouse gas but it does not harm people's lungs.

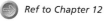 *Ref to Chapter 12*

Pollutants from engines and other sources

Engines pollute the air with harmful gases for three main reasons:

- they do not burn the fuel completely,
- the fuel is not pure,
- they run at such high pressures and temperatures that nitrogen and oxygen in the air can react together to form oxides of nitrogen.

Questions

1 With the help of Figure 11.2, give examples to show that you understand what these terms mean:

- element,
- compound,
- hydrocarbon,
- chemical reaction,
- combustion.

2 Put these four words into the correct sequence to summarise the working of a four-stroke engine: bang, blow, squeeze, suck (Figure 11.3).

Figure 11.3

Schematic diagram of the four-stroke cycle of a petrol engine.

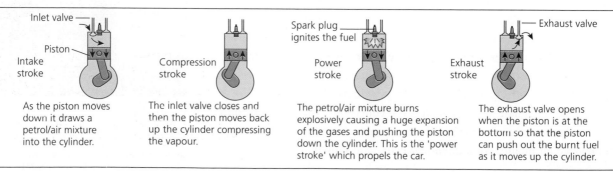

Inlet valve

Piston

Intake stroke

As the piston moves down it draws a petrol/air mixture into the cylinder.

Compression stroke

The inlet valve closes and then the piston moves back up the cylinder compressing the vapour.

Spark plug ignites the fuel

Power stroke

The petrol/air mixture burns explosively causing a huge expansion of the gases and pushing the piston down the cylinder. This is the 'power stroke' which propels the car.

Exhaust valve

Exhaust stroke

The exhaust valve opens when the piston is at the bottom so that the piston can push out the burnt fuel as it moves up the cylinder.

Pollutant	Main sources	Properties	Range of concentrations in the air we breathe
Carbon monoxide.	Incomplete combustion of fuel when there is not enough oxygen which is often the case in engines.	Colourless gas with no taste or smell which is highly toxic because it combines very strongly with haemoglobin in the blood. At low doses puts a strain on the heart and circulation. Fatal at high doses.	In the range of 10–500 p.p.b. depending on the density of traffic and the weather. WHO limit is 25 p.p.b. for exposures over 1 hour.
Oxides of nitrogen, NO and NO_2, which together are often referred to as NO_x.	Mainly from engines and furnaces which burn fuels at a high enough temperature for oxygen and nitrogen to combine.	Gases which oxidise in moist air to form nitric acid. Can make the symptoms of asthma worse.	10–45 p.p.b. but can rise to as much as 400 p.p.b. at the rush hour peak in city traffic. EU limit for NO_2 is 105 p.p.b. for exposures over 1 hour.
Sulfur dioxide.	Burning fuels which contain sulfur in motor vehicles and power stations. Main source is burning coal. There is more sulfur in diesel fuel than in petrol.	Colourless acidic gas which oxidises in moist air to sulphuric acid. Makes the symptoms of asthma worse and causes wheezing and bronchitis.	Normally well below 35 p.p.b. in urban areas but can rise to 500 p.p.b. or more. WHO limit is 122 p.p.b. averaged over 1 hour.
Ozone.	Formed when sunlight shines on polluted air containing oxides of nitrogen.	Highly reactive form of oxygen with O_3 instead of the normal O_2 molecules. Irritates the eyes, nose and throat. Can make asthma worse.	Background levels are usually below 15 p.p.b. but in summer and in the south of England levels can rise to 60 p.p.b. or more. EU guide figure is 90 p.p.b. over 1 hour and the threshold for public warnings is 180 p.p.b.
Particulates.	The coarser particles include dust from roads and industry. Finer particles include soot (carbon) from diesel engines. Sulfur dioxide and NO_x react to form fine particles of sulfate and nitrate salts.	Of great concern are the PM_{10} particles with such small size (less than 10 μm) that they can penetrate deep into people's lungs. Diesel engines are the main source of PM_{10} particles in towns and cities.	The levels of pollution recorded depend on the method of measurement. In urban areas levels range from 10–150 μg m^{-3}. The WHO health guideline for PM_{10} is 70 μg m^{-3} averaged over 24 hours.
Volatile organic compounds (VOCs) which are normally gases or liquids but evaporate so easily that their vapours stay mixed with the air.	VOCs include wide variety of carbon compounds which are mainly hydrocarbons from the evaporation of fuel and from unburnt fuel out of engine exhausts. Another source is leakage from pipelines and storage tanks.	The main VOC is usually methane from natural gas. People are worried about benzene which makes up about 2% of petrol and can form in engines as fuels burn. Benzene can cause cancer (it is a carcinogen).	Measures often distinguish methane from other non-methane volatile organic compounds (NMVOC). The total NMVOC concentration normally reaches a concentration of several hundred p.p.b. in urban air.

Concentration units

p.p.b. = parts per billion / p.p.m. = parts per million / μg m^{-3} = micrograms per cubic metre (a microgram is a millionth of a gram)

mg m^{-3} = milligrams per cubic metre (a milligram is a thousandth of a gram)

Questions

3

a The mass of a litre of water is 1 kg. Show that a concentration of 1 mg per litre is the same as 1 ppm.

b What does 1 mg of salt look like? Could you taste a salt solution containing 1 mg per litre?

4 Give examples of pollutants produced by car engines because:

a they do not burn their fuel completely,

b the fuel is not pure,

c they run at high pressures and temperatures.

5

a Why do car engines produce nitrogen dioxide even though there is no nitrogen in petrol?

b Why is it important for a car engine to be adjusted so that the air/fuel ratio supplied to the cylinders is correct?

Ozone and its effects

Figure 11.5

Photochemical smog.

Ozone often features in stories about environmental problems. The ozone layer in the upper atmosphere is a 'good thing'. The ozone high in the stratosphere protects living things by absorbing harmful UV radiation from the Sun. CFCs and other pollutants, such as the pesticide methyl bromide, tend to destroy the protective ozone layer.

In the lower atmosphere ozone is a 'bad thing' because it is harmful to living things. Ozone is a very reactive gas which irritates eyes and causes breathing difficulties. Ozone is toxic and can affect the growth of crops as well as other plants including trees. The gas is reactive enough to attack materials such as fabrics and rubber which gradually disintegrate if exposed to the gas.

Most of the pollutants shown in Figure 11.4 are primary pollutants. Ozone is a secondary pollutant which forms as sunlight shines on air containing nitrogen dioxide. The energy from sunlight splits up nitrogen dioxide molecules, NO_2, into nitrogen monoxide molecules, NO, and oxygen atoms, O. The free oxygen atoms then link up with oxygen molecules, O_2, to make ozone, O_3.

In still, sunny weather near cities the levels of ozone rise and mix with unburnt hydrocarbons from motor vehicles (VOCs, see Figure 11.3) converting some of them to compounds which together create a soup of eyewatering, irritant chemicals. A yellow haze appears. The action of sunlight on chemicals thus creates photochemical smog (Figure 11.5).

How do scientists monitor air quality?

The monitoring network

There are over 1500 hundred sites in the UK where scientists measure air quality. The sites vary in the range of measurements and methods. The aims of monitoring are to:

- understand the scale of air pollution so that the authorities can devise cost-effective ways of dealing with the problems,

- observe trends to see if policies for lowering air pollution are working,

- inform the public about air quality and warn people who may be at risk.

For each kind of information and measuring method there is a network of sites which work in the same way so that they data can be compared.

Over a hundred of the sites have automatic instruments which take measurements every hour and send the data to a central data base via a modem (Figure 11.6).

Questions

6 Draw a diagram, in the style of Figure 11.2, to show how ozone forms when the Sun shines on polluted air.

7 In the UK ozone levels are higher in Summer than in Winter; they are also higher in the south than in the north. How do you account for this?

8 In the city of Los Angeles, on still sunny days, the levels of nitrogen dioxide and hydrocarbons reach a peak at 8 am. The peak formation of photochemical smog is between noon and 2 p.m. How do you account for this?

Some sites collect information less frequently using special apparatus such as diffusion tubes or filters to collect samples for chemical analysis (Figure 11.7). Over 300 local authorities help with a network which provides a picture of levels of nitrogen dioxide pollution across the UK. Each authority arranges to place diffusion tubes at at least four sites: one within 1–5 metres of a major road, one intermediate site 20–30 metres from a main road and two background sites in residential areas more than 50 metres from a major road.

Another approach to monitoring air quality is to make estimates of the emissions from the main sources such as power stations, industry and from major roads. The estimates are collated on maps showing the main pollutants released into the air in each 1 km x 1 km square. This approach is necessary because the monitoring network gives poor coverage of the country as a whole.

Scientists are now giving much more attention to particulates which are less than 10 µm across (PM_{10}s) (Figure 11.8). These are the particulates which people are likely to breathe in and inhale deep into their lungs.

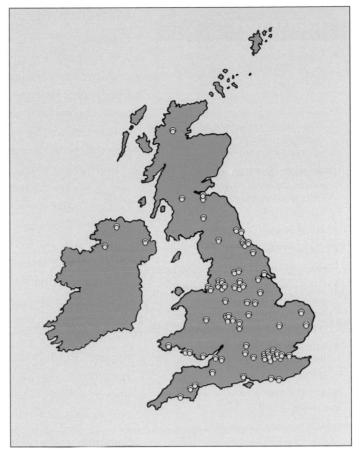

Figure 11.6

The network of automatic air quality monitoring sites in 1998. There were 89 urban sites and 19 rural sites. The instruments measure the level of pollution by oxides of nitrogen, hydrocarbons, carbon monoxide, sulfur dioxide, ozone and particulates.

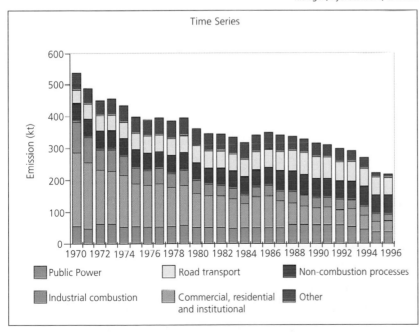

Time Series

Emission (kt)

Public Power — Road transport — Non-combustion processes
Industrial combustion — Commercial, residential and institutional — Other

Figure 11.8

Emissions of PM_{10}s from a range of sources

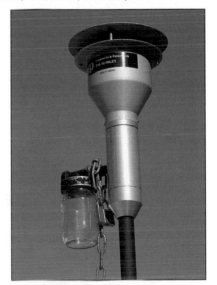

Figure 11.7

A particulate analyser used for measuring the concentration of pollutant particles in a sample of air.

9 Suggest reasons to account for the distribution of monitoring sites in the UK (Figure 11.6).

10 Why do you think that the number of monitoring sites has risen rapidly and continues to rise from under 20 sites in 1986 to over a 100 sites by 1998?

11 Where in the UK would you expect the levels of these pollutants to be particularly high or especially low:

- sulfur dioxide,

- nitrogen dioxide,

- carbon monoxide?

12 Which sectors have contributed most to the decline in PM_{10} emissions since 1970? Suggest reasons for the main trends.

13 Estimate the proportion of total PM_{10} emissions from road transport:

a in 1970, and

b in 1996.

Measurements and their accuracy

Errors

Any scientific measurement is affected by errors which are not mistakes but unavoidable differences between measured values and the true values. We can never be sure we know the true value of anything but it is possible to assess how close to the true value a measurement is likely to be. So analysts measuring air pollution can assess the uncertainty in their measurements.

There are random errors which cause repeat measurements to vary and to scatter around a mean value. Averaging a number of readings helps to take care of random errors. The average value is the best estimate of the true value.

There may also be systematic errors which affect all measurements in the same way, making them all lower or higher than the true value. Systematic errors do not average out. Identifying and eliminating systematic errors is important for increasing the accuracy of data. Some systematic errors can be corrected by checking the reading with another instrument. If both agree, this increases confidence in its validity. But it is still possible that something in the measurement procedure makes all measurements high or low. The better our understanding of the measurement process, the more it is likely to be able to spot, and remove, any causes of systematic error.

Sampling and analysis

Taking a sample is a vital first step in any measurement of air quality. It is important to try to make sure that the sample is representative of the air in the place where pollution is being monitored.

Much depends on where the sample is taken. One of London's air quality monitoring stations stands in Russell Square. The roads round the square are busy with traffic all day and much of the night but the monitoring station stands about 25 metres from the nearest road in a part of the square which is grassy and surrounded by trees and shrubs which protect the gardens from the full impact of the exhaust fumes. Moving the monitoring station to the road side would certainly affect the readings. Even changing the height at which the sample is collected would have an effect.

Another factor affecting the measurements is the sampling method. One common method is to use a pump to draw air continuously through an analyser. This method can introduce a range of errors. For example, some of the

Figure 11.9

An employee reads an air quality monitoring printout at Pacific Gas & Electric's Technical and Ecological Services Centre in Contra Costa County, California USA.

pollutants in the air may be absorbed onto the walls of the tubes as it flows through to the measuring instrument. Alternatively some of the pollutant gases may react with each other once they enter the apparatus.

Another approach to sampling is to grab a sample of the air in a bag, syringe or evacuated bottle and then to take the sample to a laboratory for analysis. It is clearly very important that the containers are clean, free of leaks and made of an inert material. Otherwise, the sample will change composition between the time of collection and the time of analysis.

Given a sample, the analyst next has to choose a method of measuring the concentration of the pollutant which is specific to that pollutant, sensitive enough to cover the range of concentrations expected and reliable.

Uncertainty in the results

Analysts have a variety of techniques for checking the validity of their results. One method is to calibrate the instruments with samples prepared with a known concentration of the pollutant. Another technique is to collect several samples in the same place at the same time and to study the variation in the results.

The automatic and urban monitoring network run by the Department of the Environment aims to achieve results in which the range of uncertainty is ± 5 p.p.b. for sulfur dioxide, nitrogen dioxide and ozone and ± 5 µg m^{-3} for PM_{10}.

How do air pollutants affect human health?

It is one thing to measure the levels of pollution. It is quite another to determine the extent to which individuals are exposed to polluted air. On a day when air quality is poor, someone living in the suburbs and spending most of the day indoors will experience very different levels of pollutants from a commuter travelling into the city centre and walking along streets jammed with traffic.

There are big research programmes in progress in many countries to try to establish the effects on health of air pollution and specific pollutants. Many people think that the steady rise in cases of asthma is connected with pollution from motor vehicles but so far there is not sufficient evidence to establish that there is a connection (Figure 11.10).

Nevertheless it is clear that asthmatics and people with lung diseases such as bronchitis are particularly vulnerable to poor air

Questions

14

a Why is it impossible to be sure that a single measurement of the concentration of nitrogen dioxide in a busy street gives a true value of the level of NO$_2$ pollution in the air?

b What steps can scientists take to increase their confidence that the value they quote for the concentration of nitrogen dioxide in the air is close to the true value?

15 Which would you expect to introduce the bigger errors into the measurement of air pollution: the methods of sampling or the methods of analysis? Give your reasons.

16 Estimates of PM$_{10}$ emissions are improving but the level of uncertainty is still high. Estimates of PM$_{10}$ emissions cover particles from diesel engines, from power stations, from industrial processes and from mining, quarrying and construction. Which of these estimates would you expect to be most reliable? Which would you expect to be least reliable?

Figure 11.10

The weekly incidence of asthma (averaged over 12 week periods) as recorded by GPs.

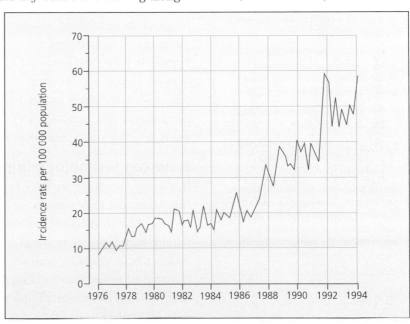

17 Study the two graphs in figure 11.11.

a Which graph shows a stronger correlation between the relative death rate and the concentration of particulates?

b What conclusions can you draw from the two graphs?

18

a What would you say to someone who suggested that the small amount of pollution from car engines is not a problem because it quickly disappears into the atmosphere?

b Does the pollution released into the air eventually disappear?

19 What steps can an individual take to protect themselves from the effects of air pollution?

Figure 11.11

Death rates in six US cities plotted against (a) the PM_{10} concentration and (b) the $PM_{2.5}$ concentration. $PM_{2.5}$ particulates are very fine being less than 2.5 μm across.

Discussion point

Is it true that most people benefit from widespread use of motor vehicles while only a minority suffer the ill effects? What are the implications in a society in which the many can outvote the few?

Is all air pollution the result of human activity or can there be 'natural' air pollution? Are these examples of pollution:

- the sulfur dioxide released into the air by volcanoes,
- the oxides of nitrogen released into the air by denitrifying bacteria in the soil?

Across the world the natural sources of nitrogen oxides give out much more NO_x than vehicles and power stations. So why do we have to worry about the human additions of these gases to the air?

quality and that today it is motor vehicles which are mainly responsible for the pollutants that are harmful to health.

In 1993, scientists from Harvard published the results of a study which began in the 1970s. Groups of young people from six cities were selected for comparison. Their health records were monitored for the next 20 years. At the same time the concentrations of air pollutants in the cities were measured. Figure 11.11 shows some of the results. The 'rate ratio' is a measure of the death rate of the groups studied compared with a control group and allowing for factors such as age, body weight, income and smoking.

The results in Figure 11.11 support the growing concern about the effects of particulates on health. This study showed that the levels of particulates typically found in European cities reduce the average lifespan by 2–3 years.

The main damage caused by the fine particles is probably caused by the chemicals they carry with them which include unburnt hydrocarbons and carcinogenic compounds.

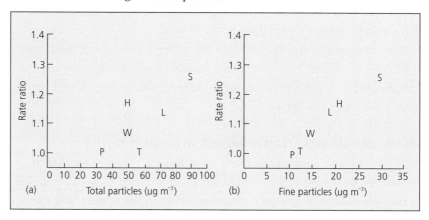

Carbon monoxide combines strongly with haemoglobin in the blood. This reduces the amount of oxygen that the blood can carry from the lungs to the tissues and as a result the heart has to work harder to pump more blood around the body. In this way air pollution by carbon monoxide puts a strain on the heart and circulation (the cardiovascular system). Exposure to 30 ppm carbon monoxide for an hour or to 9 ppm for eight hours causes headaches and affects mental performance to a measurable extent. These levels can be reached out of doors in a town on a day when the air is severely polluted.

How can technology improve air quality?

Our town planning laws have been based on motor transport for over fifty years and huge sums have been spent building roads and setting up out-of-town shopping centres. As a result our way of life now requires many people to travel a long way to work by road. This means that it would be economically very difficult to make the choice for pure air rather than unlimited motor transport. What is appropriate in towns may be inappropriate in rural areas. So there are many difficult choices to be made if pure air is to be given priority.

To date the main approach to improving air quality has been to find technical ways of changing the design of cars and the composition of fuels. As the number of vehicles on the roads of the UK doubled between 1970 and the mid-1990s the main method of cutting air pollution was to bring in catalytic converters. A regulation of the EU has required that all cars sold in Europe be fitted with catalytic converters since 1993 (Figure 11.12).

The catalyst is an alloy of platinum and rhodium. These are both very expensive metals but only very small amounts are needed. They are spread very thinly all over the surface of a ceramic block pierced with a vast network of fine holes. This gives a very large surface area of catalyst to speed up a whole series of chemical reactions which convert the harmful gases (oxides of nitrogen, unburnt hydrocarbons and carbon monoxide) to gases which are harmless to health (carbon dioxide, nitrogen and water).

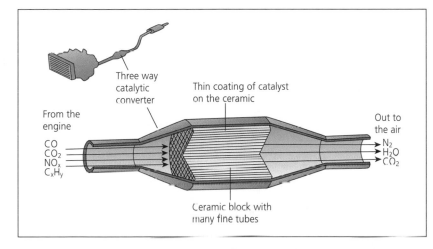

Figure 11.12

A catalytic converter

The catalyst is only effective when hot. On a short journey the catalyst may not get hot enough to start working. Also converters only work effectively if the engine is properly tuned so that the cylinders are supplied with the right mixture of air and fuel.

Catalytic converters do nothing to reduce carbon dioxide emissions. In fact they may make matters worse by slightly increasing fuel consumption. 🔘 *Ref to Chapter 12*

How can regulations improve air quality?

Why do we need controls?

Governments often impose controls when members of the public are anxious about risks to their health and safety. But there are other pressures on governments such as the need for military defence. In the 16th century coal was imported to London from Newcastle and even Queen Elizabeth was much offended by the choking smoke. Parliament was much more concerned about having enough wood for building ships for the navy to fight the Spanish. So it passed a law against cutting trees for firewood. As a result Londoners had to burn coal to cook and keep warm.

As steam power developed with the Industrial Revolution, air pollution got steadily worse. In 1819 Members of Parliament were so angry about the volume of black smoke in Whitehall that they set up a committee to 'inquire how far it may be practicable to compel persons using steam engines to erect them in a manner less prejudicial to public health and public comfort'.

During the 19th century parliament began to pass more laws to protect people by regulating working conditions and controlling pollution. An act in 1845 sought to limit harm from the smoke of railway engines and 2 years later a second act dealt with factory smoke.

Even as parliament was beginning to take cautious steps to deal with air pollution, new industries were making matters much worse. The chemical industry grew rapidly. Widnes and Runcorn, on the banks of the Mersey, were transformed from pleasant rural districts to international centres for new industries based on salt. One process for making alkali on a large scale was developed by a French chemist, Nicholas Leblanc. This process emitted huge quantities of hydrochloric acid gas into the air that devastated all the land around. It was far more acidic than the acid rain affecting the environment today. The solid waste from the process was dumped in vast heaps outside the factory where it slowly released a steady stream of poisonous and foul-smelling hydrogen sulfide gas with its intense smell of bad eggs. Living and working conditions were appalling (Figure 11.13).

Figure 11.13

Smoke over Widnes in 1895.

Inspection and authority

In the 19th century the public began to demand action to control pollution but parliament was anxious not to put the slightest check on industrial development. Dickens mocked the attitude of the industrialists to new controls when writing *Hard Times* in 1854. He describes how they were:

> ruined when they were required to send labouring children to school; they were ruined when inspectors were appointed to look into their works; they were ruined when such inspectors considered it doubtful whether they were quite justified in chopping people up in their machinery; they were utterly undone when it was hinted that perhaps they need not make so much smoke.

In 1863 Parliament passed the first of the Alkali Acts which set up an 'Alkali Inspectorate' to travel the country to check that at least 95% of acid fumes were removed from the chimneys of factories. The inspectors were scientists so for the first time science was used to protect people and the environment. Dressed in Victorian frock coats and top hats, carrying the measuring equipment up long ladders to sample the smoke from the top of factory chimneys in all weathers, they did vitally important work.

Towards the end of the 19th century the problems of the Leblanc process were solved not by controls but by developing new methods of manufacturing alkalis that did not create air pollution and which are still in use today. Even so the Leblanc process was not phased out until the early 1920s and the new processes brought fresh problems of water pollution that were only solved in the late 20th century.

Meanwhile nothing could be done about domestic smoke because each householder produced too little to be a nuisance on its own and people did not want to be protected to the extent of losing any rights or freedoms. As a result thick smogs (smoke plus fog) were an accepted feature of life in big cities such as London.

Parliament only took decisive action after 4000 people died in the 5 days during a London smog in 1952 and then 1000 people in 3 days in 1956 (Figure 11.14). Tighter regulation became possible because cheaper oil was becoming available on a much larger scale as an alternative to coal. The first Clean Air Act was passed in 1956. This act forced householders to burn smokeless fuel in controlled zones, specified minimum heights for factory chimneys and banned the emission of dark smoke, but with some exceptions. A further act in 1968 tightened up these regulations.

Since 1970 there has been a succession of directives from the European Union. These have brought in tighter and tighter controls over the emissions from motor vehicles and industry. In 1991 the UK government introduced the Road Vehicles Regulations which set limits for emissions of carbon dioxide and hydrocarbons to be included in the MOT (Ministry of Transport) test for vehicles with petrol engines.

Particularly significant was the 1995 Environment Act which created for the first time a statutory framework for local management of air quality and required the government to set air quality standards. Two years later saw the publication of the final version of the National Air Quality Strategy with targets to be met by 2005.

Figure 11.14

The big crawl. Londoners struggling home through the smog in December 1962. The smoke concentration was estimated to be nine times the average but still far below the level of the killer smogs of the 1950s.

An important aim of the National Air Quality Strategy is to cut the number of times that air pollution is so serious that it is labelled as an 'episode'. The strategy sets targets for cutting the emissions of eight pollutants which have serious effects on health. Local authorities are responsible for gathering the data they need to assess air quality in their area and make sure that the targets are met by 2005.

Review Questions

23 The aims of the 1995 Environment Act will be achieved by applying two principles:

- Best Available Technique Not Entailing Excessive Cost (BATNEEC);
- Best Practicable Environmental Option (BPEO).

a What do you understand by these two principles?

b Suggest reasons why it may prove difficult to apply these principles in practice.

c Why might a government minister, a car manufacturer and an environmental campaigner have very different views about the implications of these principles?

24 What are the scientific and technical uncertainties which make it hard to know what to do about air pollution?

25 What are the social, political and economic difficulties which make it hard to tackle air pollution problems effectively?

26 'Cutting down pollution from cars is easy, any driver can do it. All you have to do is drive less'.

- Is this a realistic way of cutting air pollution?
- What restrictions on the use of private cars to you expect to see coming in during your life time?

27 'The polluter should pay'. How might this principle apply to motor vehicles? Who should pay:

- drivers of private cars?
- passengers on public transport?

28 Since poor air quality is harmful to health would it be right to have regulations which require zero emissions of harmful gases?

29 Who cares about air pollution? Do you care? What can individual citizens do? What is the role of pressure groups? Can they make a difference? Can local authorities or governments be said to care? Is legislation enough?

Chapter 12

Fuels and the global environment

The issues

Although very real to people in some island nations (Figure 12.1), the threat of global warming is remote to people in industrialised countries. This makes it hard to make political changes, despite the view of most scientists that the consequences of global warming are serious. Some big oil companies oppose change fearing that profits will be hit by new regulations. Other industries do not want to bear the cost of reducing emissions by changing how they do business. Less developed countries are unhappy about anything that might limit their opportunities to industrialise.

The science behind the issues

Burning fossil fuels has done more than anything else to increase the concentration of greenhouse gases in the atmosphere. When fuels burn the chemicals in them do not simply disappear. The products of burning, which pour out of chimneys and exhaust pipes, may be changing the climate.

The Earth's climate and life on Earth depend on radiation from the Sun which is 150 million kilometres away. The greenhouse effect and global warming are a consequence of the absorption of radiation from a distant source.

Figure 12.1

Rising sea levels caused by global warming are a likely threat to people living on low-lying land. The people of Kiribati are watching their islands submerge as the tides get higher. Most of the 33 small coral atolls which make up Kiribati are no more than 2 metres above sea level. Rising sea levels mean that salt water contaminates their fields and their supplies of drinking water.

What this tells us about science and society

Scientists can only gather a limited amount of data. They know that the level of carbon dioxide in the atmosphere is rising but there is still a debate about the extent of global warming. There has always been natural variation in our climate so it is hard to tell if a change would have happened anyway.

Predicting changes in the climate over the next 10 or 20 years is much more uncertain than trying to produce a weather forecast for the next few days. Even so, scientists are gathering data and feeding them into their computer models. These models are only as reliable as the assumptions on which they are based and the quality of the data. Most scientists, however, are now convinced that global warming is a serious threat.

There is usually moisture (water vapour) in the air. **Dry air** means air with no moisture in it. Humidity is a measure of the level of water vapour in the air.

Low **concentrations** of pollutants are often measured in parts per million, p.p.m. – by mass or by volume. The proportion of carbon dioxide in dry air is 0.0364 per cent by volume which is 364 ppmv.

Reservoirs are parts of the Earth system such as the atmosphere, the oceans or the land. Reservoirs are stores of elements or compounds. The atmosphere consists mainly of the elements nitrogen and oxygen but also contains carbon in the form of the compound carbon dioxide.

A **flux** is a flow.

Figure 12.2

Different natural forms of carbon as an element or as compounds. Carbon is one of the hundred or so elements which combine to make all the natural and synthetic chemicals. What is special about carbon is that its atoms can join up in many different ways in combination with a few other elements such as oxygen, nitrogen and phosphorus and this gives rise to all the many chemicals in living things including carbohydrates, proteins, fats and DNA.

Does human activity really make a difference?

The carbon cycle

The atmosphere consists mainly of four gases: nitrogen, oxygen, argon and carbon dioxide together with varying amounts of water vapour. Carbon dioxide makes up only 0.0364% of the volume of dry air.

Carbon dioxide is released into the air naturally by respiration, by the decay of living things after death and from forest fires. Photosynthesis in plants takes in carbon dioxide. Human activity increases the proportion of carbon dioxide in the air by burning fossil fuels and by deforestation.

During the past 200 years, since the start of the industrial revolution, the concentration of carbon dioxide in the air has risen from about 0.0280% to 0.0364% of the volume of the air. The figures look small but the increase is significant.

A useful way of describing what happens to carbon in the environment is to draw a diagram showing the main reservoirs of the element with arrows to show the flows between them (Figure 12.2). This is called the carbon cycle.

Diamond and graphite consist of pure carbon. They are made up only of carbon atoms

Plants turn carbon dioxide from the air and water into sugar. Sugar consists of carbon combined with hydrogen and oxygen

Animals breathe out more carbon dioxide then they breathe in. Carbon dioxide consists of small molecules (CO_2)

Chalk, limestone and marble are all made of calcium carbonate, a compound of calcium, carbon and oxygen

A series of chemical changes means that the carbon appears in different forms in the different reservoirs but it does not disappear at any stage. The total quantity of carbon on the Earth is constant.

In fossil fuels the carbon is combined with hydrogen. The simplest example is methane in natural gas. Another simple carbon compound is carbon dioxide and that is the form the element takes up as a gas in the air. In living things carbon is present in almost all the important molecules such as carbohydrates, proteins and fats as well as the molecules of genetics such as DNA.

Watching the build-up of carbon dioxide

Scientists have been measuring the carbon dioxide in the air for many years. They face a formidable challenge. The proportion of carbon dioxide in the air is small and the changes in concentration each year are even smaller.

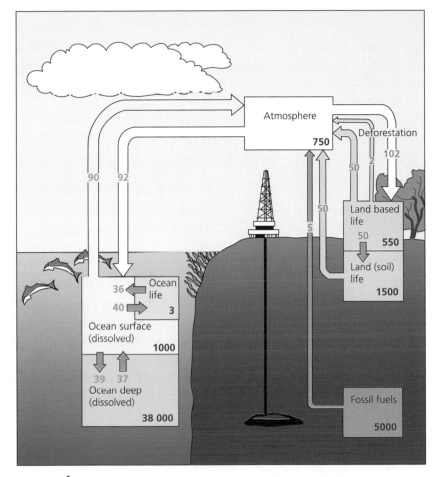

Figure 12.3

The global carbon cycle. The boxes are reservoirs of the element carbon in different combinations with other elements. The red numbers are estimates for the total mass of carbon (worldwide) in the reservoirs in gigatonnes. The green numbers are fluxes showing the rate of movement of carbon from one reservoir to another in gigatonnes per year. (One gigatonne = 1000 million tonnes.)

Questions

1

a According to Figure 12.3, what is the total mass of carbon passing into the atmosphere each year?

b What is the total mass of carbon passing out of the atmosphere each year?

c What is the net change in the mass of carbon in the atmosphere per year?

d Do the figures in the diagram suggest that human activity is having a significant effect on the global carbon cycle?

2 Comment on the meaning of the term 'carbon cycle'. What is the value of thinking of what happens to carbon in the environment in terms of a cycle? Is there just one cycle or are there several cycles?

3 The values in Figure 12.3 are estimates.

a Explain why there is considerable uncertainty in the numbers.

b Which of the values do you think is most difficult to estimate?

c Which of the values might be estimated fairly reliably?

The technique for making the measurements is similar to one of the methods for measuring the alcohol in a driver's breath. Molecules such as carbon dioxide and alcohol absorb infrared radiation (see Figure 12.6). The greater the concentration of the molecules the stronger the absorption. The instrument is an infrared gas analyser which continuously records the concentration of carbon dioxide in a stream of air flowing through it.

A problem for the scientists is to find a place to make the measurements where the results will not be affected by exhaust gases from nearby traffic, or by the gas exchange of nearby crops as they photosynthesise or respire, or by industry and commerce.

The best known site for making measurements is at the Mauna Loa Observatory in Hawaii off the west coast of the US mainland. Mauna Loa is the highest mountain in Hawaii and the observatory is 3400 metres above sea level. Scientists have been using the same procedure for taking readings ever since 1958. The air samples are collected through air intakes 7 metres and 27 metres above the ground in the midst of a barren field of volcanic rock (Figure 12.4).

Question

4 Study Figure 12.4.

a What times of year correspond to the peaks and troughs in the figure? Suggest an explanation for the zig-zag pattern.

b What was the average percentage increase in the carbon dioxide concentration at Mauna Loa between 1958 and 1998?

c Why is Mauna Loa a good place for making observations of this kind?

Figure 12.4

A graph to show the measurements taken at the Mauna Loa observatory between 1958 and 1998. The 1997–98 increase in the annual growth rate is the largest single yearly jump since the record began.

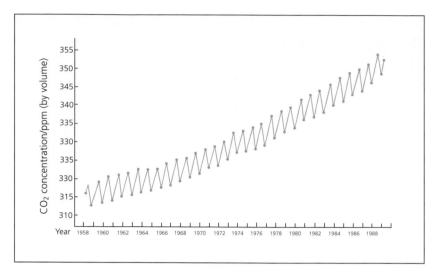

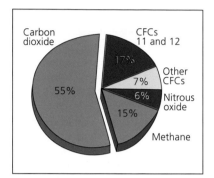

Figure 12.5

The contribution to the greenhouse effect by the gases which human activity adds to the air. Overall water vapour is the most significant greenhouse gas. The contribution to the greenhouse effect made by water is twice as much as all the effects of all gases shown in this chart.

Why does carbon dioxide affect the Earth's climate?

Radiation from the Sun (see Figure 12.5) falls on the Earth. The atmosphere reflects about 30% of it into space, gases in the air absorb another 20% and about half reaches the Earth's surface (see Figure 12.6).

The radiation from the Sun warms the Earth's surface which then radiates energy back into space but at longer, infrared wave lengths. Much of this infrared radiation is absorbed and by the atmosphere, warming it up. This is the greenhouse effect (Figure 12.7).

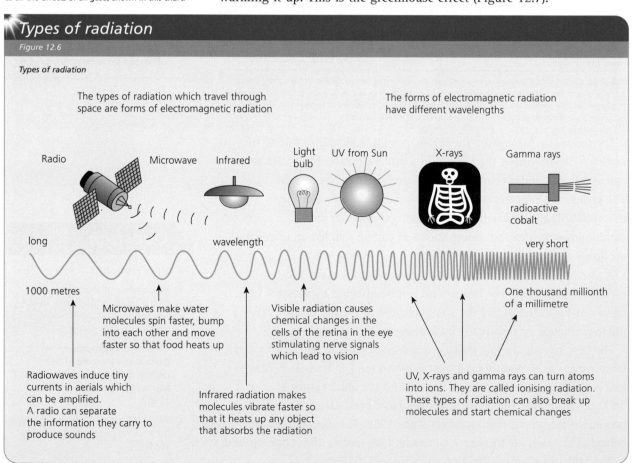

Types of radiation

Figure 12.6

Types of radiation

The types of radiation which travel through space are forms of electromagnetic radiation

The forms of electromagnetic radiation have different wavelengths

Radio   Microwave   Infrared   Light bulb   UV from Sun   X-rays   Gamma rays

radioactive cobalt

long   wavelength   very short

1000 metres

One thousand millionth of a millimetre

Radiowaves induce tiny currents in aerials which can be amplified. A radio can separate the information they carry to produce sounds

Microwaves make water molecules spin faster, bump into each other and move faster so that food heats up

Infrared radiation makes molecules vibrate faster so that it heats up any object that absorbs the radiation

Visible radiation causes chemical changes in the cells of the retina in the eye stimulating nerve signals which lead to vision

UV, X-rays and gamma rays can turn atoms into ions. They are called ionising radiation. These types of radiation can also break up molecules and start chemical changes

The greenhouse effect keeps the surface of the Earth 33 °C warmer on average than it would be if there were no atmosphere. Without the greenhouse effect there would be no life on earth. The Moon's average temperature is about −18 °C because it has no atmosphere.

The gases in the air which absorb infrared radiation are known as greenhouse gases. Nitrogen and oxygen make up most of the air but they are not greenhouse gases. The main natural greenhouse gases are carbon dioxide, methane, nitrous oxide and water vapour.

The rising concentrations of carbon dioxide and other greenhouse gases in the atmosphere are enhancing the greenhouse effect. It is this enhanced greenhouse effect which many blame for global warming and climate change.

Questions

5 Sketch a diagram to explain why the greater the distance of a source of radiation the weaker the effect of the radiation when it falls on another object.

6 Give examples from your everyday experience of radiation being:

a reflected;

b transmitted; and

c absorbed.

7 How would you explain to a younger person the extent to which we all depend on electromagnetic radiation in daily life?

8 Why does more energy from the Sun fall on each square metre of the planet Mercury than on each square metre of the Earth?

9 All hot objects radiate energy. How does the radiation given off from the surface of the Earth differ from the incoming radiation from the Sun and why is this important?

10 Why the term 'greenhouse effect'? Is the Earth really like a greenhouse?

11 Why is it important for human health and for other living things that the atmosphere absorbs most of the ultra-violet radiation from the Sun and only allows visible light and some UV and IR radiation to reach the ground?

The greenhouse effect

Figure 12.7

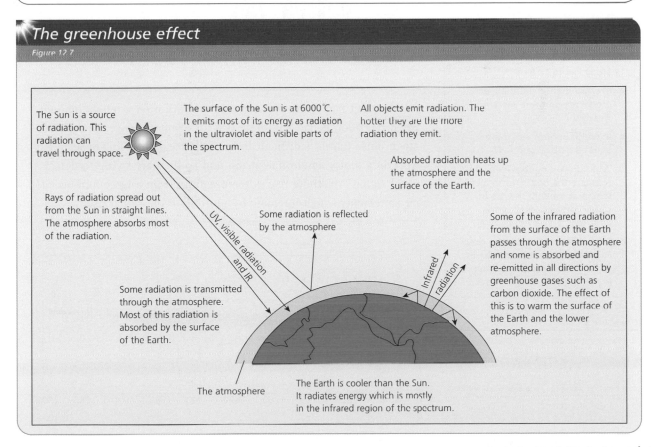

The Sun is a source of radiation. This radiation can travel through space.

The surface of the Sun is at 6000 °C. It emits most of its energy as radiation in the ultraviolet and visible parts of the spectrum.

All objects emit radiation. The hotter they are the more radiation they emit.

Absorbed radiation heats up the atmosphere and the surface of the Earth.

Rays of radiation spread out from the Sun in straight lines. The atmosphere absorbs most of the radiation.

UV, visible radiation and IR

Some radiation is reflected by the atmosphere

Infrared radiation

Some of the infrared radiation from the surface of the Earth passes through the atmosphere and some is absorbed and re-emitted in all directions by greenhouse gases such as carbon dioxide. The effect of this is to warm the surface of the Earth and the lower atmosphere.

Some radiation is transmitted through the atmosphere. Most of this radiation is absorbed by the surface of the Earth.

The atmosphere

The Earth is cooler than the Sun. It radiates energy which is mostly in the infrared region of the spectrum.

Is the Earth warming up?

Evidence from measurements

There is no doubt that the concentration of carbon dioxide in the air has risen. The theory of the greenhouse effect shows that more carbon dioxide in the air should lead to global warming. Climate scientists, however, have not been able to show definitely that global temperatures have risen as a result – though the evidence that it has is getting stronger.

The problem is that there are lots of other reasons for variations in climate. The causes of natural climate changes include volcanic eruptions, variations in the amount of energy reaching the Earth from the Sun, fluctuations in ocean currents as well as large flows of hot or cold air in the atmosphere of the Earth itself.

There is enough evidence from the past to show that even before the industrial revolution the Earth's climate varied by as much or more than the 0.5 °C rise in temperature observed over the last hundred years or so (see Figure 12.8).

Figure 12.8

The changes in the mean global temperature since 1872. The red, rising line shows the mean values of measurements of the temperature at the Earth's surface. The purple bars show how the recorded values year by year fluctuate about the trend line.

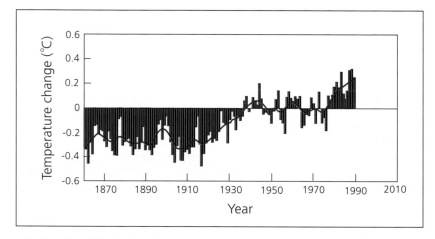

Questions

12 Study Figure 12.8.

a In which period did the mean global temperature remain steady?

b In which periods was there marked global warming?

c What do you notice when you compare the rise in mean global temperature with the rise in carbon dioxide levels in the atmosphere (see Figure 12.4)?

d Does there appear to be a close correlation between the rise in carbon dioxide levels and the rise in mean global temperatures?

It is rather like trying to have a conversation on a mobile phone when reception is bad and there is a lot of interference. There is so much noise that it is difficult to hear the message. Climate scientists work in a 'noisy' environment too and so they are trying to detect a small signal (a definite rise in temperature) from a large number of measurements which fluctuate by more than the signal itself.

Figure 12.9

A smoothed curve (blue and red line) showing a rise in mean temperature in central England against a background (grey lines) of annual fluctuations. The smoothed curve averages the data in a way that shows the variations in a timescale of 100 years.

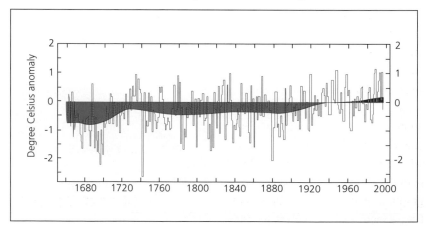

Figure 12.9 suggests a warming of about 0.7 °C over 300 years of which about 0.5 °C occurred during the 20th century. The data show that the decade from 1988 to 1997 was the warmest in the entire period with five of the warmest years since 1659.

The table in Figure 12.10 shows that the 1990s were unusually warm. To be counted as exceptional the month, season or year must be one of the ten warmest since the records began in 1659. If the variations were purely random one would only expect four entries in this table whereas there are 25. Only one month was exceptionally cold in the same period.

Questions

13 What measurements would you make if you wanted to find out whether or not the mean temperature is rising where you live? What difficulties would you expect to have in showing that there really has been a rise in mean temperature over a period of time?

14 Give examples of the checks that scientists make when trying to make sure that they can trust the values from their measurements.

Figure 12.10 Exceptionally warm months, seasons and years during the 1990s based on data collected in central England. Anomalies are shown in °C compared to the 1961–90 mean value.

January	1990 +2.7		
February	1990 +3.5	1998 +3.4	
March	1997 +2.7	1990 +2.6	1991 +2.2
April			
May	1992 +2.4		
June			
July	1995 +2.5		
August	1995 +3.4	1997 +3.1	1990 +2.2
September			
October	1995 +2.3	1990 +1.3	
November	1994 +3.6	1997 +2.0	
December			
Winter	1990 +2.2	1998 +2.0	
Spring	1992 +1.7	1997 +1.4	1990 +1.4
Summer	1995 +2.0		
Autumn	1995 +1.2		
Annual	1990 +1.15	1997 +1.06	1995 +1.05

Computer modelling and climate change

Scientists are generally much more confident about their theories if they can devise a model which will allow them to make predictions which they can compare with their observations or experimental results. The Earth's atmosphere is so complex that it only became possible to model the climate once powerful computers were developed during the 1980s.

Figure 12.11 shows how the temperature of Central England has varied since 1660 in comparison with the mean temperature in the period 1961–90. At the right of the graph, in colour, are simulations of future temperature changes based

Figure 12.11

The black curve shows the mean annual temperature of Central England from 1659 to 1997 compared to the mean temperature in the period 1961 to 1990. The red and purple curves are based on a computer model. The red lines assume high emissions of carbon dioxide. The purple lines assume low emissions.

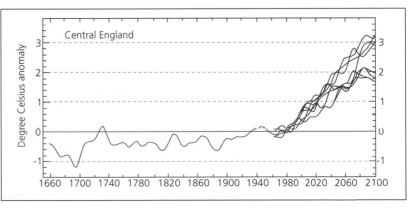

Key terms

Weather describes the state of the atmosphere such as its temperature, rainfall, wind speed and humidity at a particular place and time.

Climate describes the general patterns in the atmosphere over larger regions and longer time scales.

Questions

15 What do you understand by the '± 0.1' in the statement that overall warming in the 20th century was 0.5 ± 0.1 °C?

16 Why is there often a poor match between the observed figures for global warming and the figures from computer models which attempt to explain the temperature rise?

17 Why is it difficult to determine whether the observed rise in mean global temperature in the 20th century was due to the enhanced greenhouse effect?

Key term

Feedback affects a system when one of its properties or outputs influences the inputs which determine how it behaves. Positive feedback makes a system unstable because it can lead to a runaway snowball effect. A familiar example is the loud shriek when a microphone is too close to the loudspeaker. Negative feedback tends to be stabilising. A rising output is fed back in some way such that the input to the system is lower than it otherwise would be.

on computer models. The red lines are simulations based on the assumption that carbon dioxide emissions world-wide will not be controlled but will continue to rise. The purple lines indicate the likely rise in mean temperature if there are cuts in the use of energy and the emissions of carbon dioxide are lower.

Meteorologists have developed sophisticated computer models for weather forecasting and it is from this work that they have developed their climate models. They base these models on the known laws of physics which they use to write mathematical equations to describe what happens to the gases in the air, to water vapour and to the energy from the Sun. The aim is to explain climate changes in the past and to predict how the climate will change in future in response to variations in the levels of greenhouse gases.

The atmosphere is so large and complex that climate modellers have to make big simplifications. The model used to produce the projections in Figure 12.9, for example, operates by solving a set of mathematical equations for a number of points which are about 300 km apart on an imaginary grid over the UK. This is a very coarse grid covering the UK with just four grid boxes.

Modellers have to make many assumptions when they set up their models to explore future climate change. In particular they have to decide what the rate of greenhouse gas emissions will be, based on expectations for population growth, energy use and changes in technology. Typically they explore a range of possible scenarios some of which assume high outputs of carbon dioxide while others assume that government regulation will be effective so that emissions will be controlled and cut.

Estimating the effect of increasing levels of greenhouse gases in the air is made even more complex by the impacts of global warming which themselves affect the climate. If sea ice and snow melt, for example, the amount of reflected sunlight decreases. This means that more of the Sun's energy is absorbed and warming increases even more. This is an example of positive feedback.

The modellers are most uncertain about the effects of clouds in a warmer world. Clouds reflect sunlight and so more clouds could decrease global warming – negative feedback. However, moisture in the air and clouds is a powerful greenhouse gas which traps more of the reflected radiation and enhances the warming effect – another example of positive feedback.

All this means that the results of climate modelling have to be interpreted with great caution. Nevertheless work with different models in various parts of the world suggests that:

- the average global warming will be about 0.3 °C per decade at current rates of greenhouse gas emissions;

- natural variability in climate leads to fluctuations which are also about 0.3 °C in 10 years;

- there will be regional variations with more warming in the parts of the world away from the equator and closer to the poles.

What are the possible impacts of global warming?

Global warming is expected to raise sea levels, change patterns of rainfall and, in various ways, have effects which are a threat to human health, food production, water resources and the ecosystems that sustain biodiversity.

Sea level change

Sea levels will rise in a warmer world because water expands as it gets hotter and because ice sheets on land in Greenland and Antarctica will melt. Sea levels rose by about 10–20 cm during the last century. It seems likely that much of this rise has been caused by global warming.

The rise of sea levels is already a threat to low lying islands such as Kiribati (Figure 12.1) and the Maldives, and to coastal regions in many countries including Bangladesh. A rise may only be 20–80 centimetres but this would be enough to flood very large areas of land where millions now live and grow their food.

Figure 12.12

Changes in the extent of ice shelves in Antarctica may be a sign of a serious impact of global warming. Here scientists are preparing to measure the tilt of the Ronne ice shelf by burying a tiltmeter in a trench and covering it with ice blocks to protect it from temperature changes. The instrument can measure the tiny changes in the tilt of the ice sheet caused by tides or changes in buoyancy as large icebergs break away.

Weather patterns

Thanks to the gulf stream which carries warm sea water across the Atlantic from the seas round the Caribbean islands, the climate in the UK is much warmer than it would otherwise be. Sea water stores a huge amount of energy and ocean currents distribute this energy about the globe. The gulf stream is like a conveyor belt bringing energy from the sunnier Caribbean to warm the seas round the UK. There is a theory that global warming might shift the gulf stream or alter its strength and this could mean that we would have much colder winters while being part of a generally warmer world.

Figure 12.13

A hurricane at Padaram harbour, South Massachusetts, in August 1991 where the high winds destroyed 90 boats and the harbour bridge was washed away.

Questions

18 Melting floating ice in the Arctic will not affect sea levels. Why not?

19 Suggest reasons why it is very difficult to measure past and present changes to sea levels.

20 Suggest reasons why an increase in extreme weather events could lead to more deaths from cholera and from malnutrition.

21 In what ways might human societies have to adapt to cope with the health risks in a warmer world? What are the implications of the changes?

22 What effects might global warming have on the availability of safe drinking water?

Discussion points

Do you think that the impacts of global warming are likely to be serious enough to justify the huge cost of curbing the output of greenhouse gases?

Why is it that the poorer people in the less industrialised countries are likely to suffer most from the impacts of global warming and climate change?

What, if anything, could and should be done to reduce the rate of global warming and to cope with its effects?

Key terms

An **ecosystem** is the community of plants and animals found in a particular habitat. An ecosystem can be as small as a pond or as large as a prairie or desert.

Biodiversity describes the number, range and variety of species in a habitat and includes the extent of genetic diversity within species.

Human health

Many scientists think that global warming will bring with it many more extreme weather events such as violent storms and prolonged heat waves. These can have a direct effect on human health, especially in vulnerable communities which are already poor. Meanwhile indirect effects will include a change in the distribution of the insects and micro-organisms which carry disease. At the moment about 45% of the population of the world live in regions where mosquitoes carry malaria. In a warmer world this might rise to around 60%.

Food production

Uncertainties in the climate models make it very hard to predict the impact of global warming on agriculture. In any region farmers grow the crops which suit the local climate. Changes in the climate will alter both the productivity of their crops and the nature of the crops they can grow. This would bring gains and losses. In the UK, for example, it would only take a small rise in temperature for much larger areas to grow and ripen corn (corn on the cob). A warmer climate, however, could allow pests to flourish which currently cannot survive British winters.

Plants take in carbon dioxide for photosynthesis so higher CO_2 levels in the atmosphere might mean faster crop growth for wheat and rice but they would have little effect on other food crops such as maize and sugar cane which are less responsive to higher levels of the gas.

Plants need water and so production could be much reduced in regions where the changing climate cuts rainfall. Although there will be areas of drought, a warmer world will also be a wetter world.

Ecosystems

There is a serious risk that the rate of change of climate will be so great in many regions that ecosystems will not be able to respond fast enough to survive. Ecologist generally expect ecosystems to be able to respond to temperature changes at a rate of about 0.1 °C every 10 years, but the computer models are suggesting changes that are at least twice this rate. This could mean that many regions which are important to biodiversity would be under serious threat such as forests, wetlands, coral reefs (Figure 12.14).

The global warming debate

An international panel of scientists, called the Intergovernmental Panel on Climate Change (IPCC), has convinced many politicians that global warming is a real problem and that urgent action is required by governments to cut emissions of greenhouse gases worldwide. This conclusion is based on the work of thousands of scientists all over the world. But not all scientists agree.

At the Rio Earth Summit, in June 1992, over 160 governments signed a convention agreeing to adopt a precautionary approach to the threat of global warming. What this means is that they agreed to start

Figure 12.14

This coral reef has begun to die. What is the cause? Is it the effect of global warming? Is it the result of rising levels of dissolved carbon dioxide in sea water?

taking action to deal with the emissions of greenhouse gases despite the lack of certainty that they are as harmful as the existing scientific evidence suggests.

Later in 1997, at the Kyoto conference, politicians from many nations agreed to reduce the emission of greenhouse gases by 5.2% from their 1990 level by 2008-12. Nevertheless there has been little action since and many people are effectively ignoring the warnings from the scientists.

James Hanson is a research scientist who studies climate change. He discusses the response to the warnings from scientists in an article about the global warming debate which he wrote in 1998 for the Goddard Institute for Space Studies. He quotes the famous physicist Richard Feynman, who wrote:

> The only way to have real success in science ... is to describe the evidence very carefully without regard to the way you feel it should be. If you have a theory, you must try to explain what's good about it and what's bad about it equally. In science you learn a kind of standard of integrity and honesty.

James Hanson argues that scientists, in general, are not making a good job of the global warming debate in public. Too often they appear to be arguing from a fixed point of view. They are either trying to persuade politicians and industry that global warming is serious and needs more research or they are presenting the counter-argument that global warming is a non-issue that we do not have to worry about.

He points out that the fun in science comes from being open-minded and exploring ideas from different points of view. Scientists should be challenging each other, testing new ideas and asking lots of questions to test that evidence and explanations.

Some scientists involved in the global warming debate, however, have taken up strong positions on the issue influenced by political, environmental or religious points of view. As a result they have ceased to keep the open mindedness which is often regarded as the scientific ideal. Instead they have acted like lawyers hired to promote or defend a particular point of view.

Review Questions

23 Suppose a group of people decided to change their lifestyle to cut down the contribution they make to global climate change. Below are some of their planned changes. Which of them would make a difference if a lot of people followed their example?

a bicycling to work instead of travelling by car;

b burning wood to heat homes instead of natural gas;

c adopting a vegetarian diet;

d composting organic kitchen waste;

e using paper and cardboard instead of plastics;

f showering to keep clean instead of bathing;

g planting trees.

24 Here are two questions that you might be asked about global warming. What answers would you give?

a 'It seems to me that it would not make much difference day by day if the temperature was about 1 °C warmer. How can a temperature rise of just one or two degrees make a big impact on our environment?'

b 'People and industry will not want to pay higher taxes for fuel and electricity to cut emissions of greenhouse gases. People will not understand the scientific arguments. How can I explain to people that there really is evidence now to show that we are faced with serious climate change?'

Chapter 13

Radioactivity

The issues

The general public often has a very negative attitude towards the use of radioactive substances, particularly in large scale applications, such as nuclear power stations (Figure 13.1). There is generally less opposition to the medical use of radioactive materials (Figure 13.2).

How well founded are people's fears of the uses of radioactive materials? How can these dangers be balanced against the benefits? How great is the risk to people's health of exposure to radiation? Can the risks be reduced or eliminated altogether?

Figure 13.1

Sellafield (formerly Windscale) has been the focus for fears about the nuclear industry in the UK. The two cube-shaped buildings house the original Windscale reactors which were closed after a major fire in 1957. Part of the nuclear reprocessing plant appears on the right.

The science behind the issues

Some substances are radioactive. They emit ionising radiation. There are three main types: alpha, beta and gamma. Gamma is the most penetrating, but alpha causes more ionisation along its path. An irradiated object does not become radioactive, but ionisation can cause damage to living cells. The health risks of ionising radiation increase with the dose a person receives. In addition to the risks of irradiation, a person (or the environment) may be contaminated by a radioactive substance if it is allowed to spread from its source.

What this tells us about science and society

Although the effects of very high doses of radiation are rapid, lower doses do not have immediate effects, but rather increase a person's risk of illness in the future. Data on the effects of low doses of radiation are difficult to collect, so assessing the risk involves interpretation and judgement. There are now strict regulations on exposure to ionising radiation, often based on the 'precautionary principle': if there is any doubt, then act on the side of caution. However, each further reduction in levels is more expensive to achieve. Society has to decide what limit is acceptable. This cannot be resolved by scientific information alone.

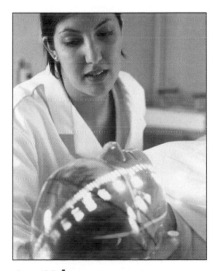

Figure 13.2

A radiographer preparing a cancer patient for radiotherapy using a laser beam to make sure that the invisible ionising radiation will be targeted at the brain tumour. The plastic mask holds the patient's head still and protects healthy cells from the radiation.

Figure 13.3

Effects of different radiation doses. For doses below 0.25 Gy there is no immediate effect, but there is a risk of cancer developing in the long term. This risk is in proportion to the radiation dose.

Key term

The unit of **radiation dose** is the **gray** (Gy). It is a measure of the amount of energy transferred by radiation to each kilogram of body tissue. (1 Gy = 1 J kg⁻¹)

Question

1

'At 10.15 a.m. on Tuesday 17th June 1997, Aleksander Zakharov made a fatal mistake. Alone behind concrete walls three feet thick, in an underground bunker ... he placed a thin shell of copper on a sphere of highly enriched uranium. Suddenly, a huge burst of radiation turned the air blue ... Zakharov ...realised immediately what had happened. He left the bunker, closed the hatch notified a manager and lost consciousness. Three days later ... he died in a Moscow hospital.'
New Scientist, 30 October 1999.

Use Figure13.3 to estimate Zakharov's absorbed radiation dose.

How well founded are the dangers of radiation?

Radioactivity in general, and the nuclear power industry in particular, are more often than not adversely reported in newspapers and in the media generally. This 'bad press' certainly contributes towards the public's fear and hostility, but is also to a considerable extent a reflection of attitudes that already exist.

Adopting a cautious attitude towards radioactive substances is entirely justified on the basis of the available evidence. There is little doubt that people exposed to high doses of radiation (for example, when the US dropped bombs on Hiroshima and Nagasaki in Japan at the end of the Second World War, or at Chernobyl in the Ukraine at the time of the serious nuclear accident in 1986) suffered dreadfully. Figure 13.3 shows the immediate effects on people in relation to the amount of radiation they received.

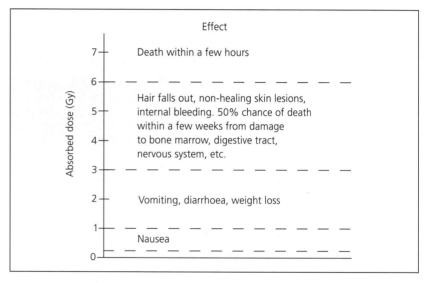

For doses below 0.25 Gy, plenty of evidence accumulated over the following years of long-term effects on people's health, mainly the greatly increased risk of various types of cancer.

Though people are right to be cautious about anything that might increase their exposure to radiation, not all of the fears concerning radiation that are common among the general public are justified. Anxieties about different sources of radiation are often completely out of proportion to the actual risk. For example, many people are far more concerned about the radiation risk of nuclear power stations than they are about the radiation risk in their own homes. In fact, however, for the great majority of people the radiation risk from their own homes is far greater than the risk from nuclear power stations. Furthermore, one very commonly held belief about radiation is incorrect – things which are exposed to radiation from radioactive substances do *not* themselves become radioactive.

- When radiation is absorbed it no longer exists as radiation.

- The energy carried by the radiation heats the material that absorbs it.

- It may also cause changes to molecules in whatever absorbs it.

Radioactivity

Figure 13.4   Figure 13.5

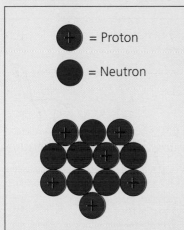

The nucleus of a carbon-12 atom. The '12' refers to the total number of nucleons (protons and neutrons).

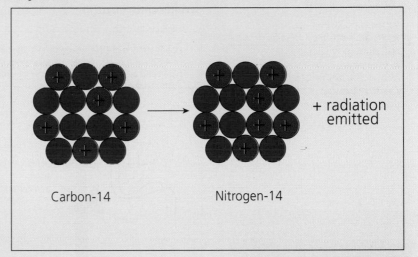

The nucleus of a carbon-14 atom emits radiation when it decays to produce the nucleus of a nitrogen-14 atom.

Carbon-14   Nitrogen-14

+ radiation emitted

Everything is made of atoms. There are about 100 different elements, each made from a particular type of atom. Atoms of different elements can, however, join together in different ways to make millions of different substances. During chemical reactions atoms are re-arranged in different ways to make new substances. But the atoms themselves do not change: there is always the same number of each type of atom at the end of a chemical reaction as there was at the start.

Most atoms are stable but the atoms of radioactive substances are unstable. They decay to produce different atoms and emit radiation as they do so. There is nothing that we can do to speed up, slow down or stop this process of decay; it just happens. Even joining radioactive atoms to other atoms to make new substances has no effect on the rate at which the unstable atoms decay. Each type of unstable atom has its own particular rate of decay.

All atoms consist of a small central nucleus containing most of the atom's mass, together with electrons (each of which has very little mass but has a negative electrical charge) in the space around the nucleus. The nucleus is made up of protons (with a positive electrical charge) and neutrons (which are electrically neutral) (Figure 13.4). Protons and neutrons have equal mass. Atoms have the same number of protons and electrons; this means that they have no overall electrical charge.

Atoms of the same element always have the same number of protons (and hence electrons), but may have different numbers of neutrons. Atoms of the same element with different numbers of neutrons are called isotopes (or nuclides) of that element. Unstable isotopes are called radio-isotopes (or radionuclides).

Radioactive decay is a change to the nucleus of an atom; it is a nuclear change. During a nuclear change an atom of one element becomes an atom of a different element (Figure 13.5).

Questions

2

a What is the difference between atoms of:

i   carbon-12 and carbon-14?

ii uranium-235 and uranium-238?

b Explain why the above pairs of atoms are isotopes of the same element.

3 What happens to the nucleus of an atom of carbon-14 when it decays to produce an atom of nitrogen-14?

Why is radiation dangerous?

When radiation is absorbed by cells, the energy carried by the radiation doesn't just heat the cells up. It can also break up molecules into fragments. Molecules are electrically neutral but the fragments into which they are broken are often electrically charged. These charged fragments are called *ions* and the radiation that produces them is called *ionising radiation*.

Key term

Ionising radiation is radiation that is able to produce **ions** (charged fragments of molecules) in a material that absorbs it.

High-intensity radiation can damage so many molecules in cells that it kills them. Cells may be able to repair lesser amounts of damage done by lower levels of radiation. If the DNA in cells is damaged, however, (see Figure 13.6) even a small amount of damage can have serious consequences.

Figure 13.6

How ionising radiation damages cells.

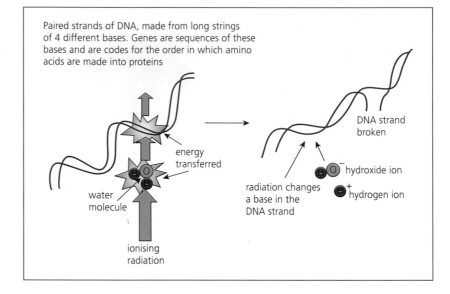

If vital genes are damaged, cells can be crippled or killed. If genes that control growth and replication are damaged, cells can start to divide uncontrollably to produce tumours of rogue cells. This is cancer. There is often a time delay of several years between the initial damage to cells and the development of a cancerous tumour. *Ref to page 53*

Because there is, for all but very low radiation doses, a good *correlation* between radiation dose and the risk of cancer and, in addition, a well established *mechanism* about how this happens, scientists are justified in claiming that ionising radiation *causes* cancer.

How dangerous are different types of radiation?

Since alpha particles are very easily absorbed by cells (see Figure 13.8) and so cause a lot of ionisation in a short distance, it might seem that they are more likely to damage cells than beta particles or gamma rays. This is certainly true if the unstable atoms that are the source of the alpha particles are close up against living cells or even inside them. However, for sources of radiation *outside* of a person's body, alpha radiation is the *least* dangerous. This is because alpha radiation from a source more than a few centimetres away from a person's body will not even reach their body.

To take account of the greater damage done by any alpha radiation that *is* absorbed by body cells, the *absorbed dose* of this type of radiation (in gray, Gy; see p.168) is scaled up by a factor 20. This is then added to the absorbed dose of beta and gamma radiation to give the *equivalent dose* (in sievert, Sv).

In addition, the different types of cell in the different tissues and organs of the body are differently affected by radiation; reproductive

Questions

4 Describe, as fully as you can, two different ways that radiation can damage the DNA in cells.

5

a Radiation is sometimes used to help cancer patients. Explain how.

b Why is this treatment risky?

c Why might this risk nevertheless be worth taking?

Three types of radiation

Figure 13.7

The electromagnetic spectrum.

high frequency   low frequency

| gamma | x-rays | ultraviolet | light | infrared | microwave | TV and radio |

ionising radiation

Though all of the radiation emitted from radioactive substances is ionising radiation, it is not all the same type. One of the types of radiation emitted is called gamma (γ) radiation. It is electromagnetic radiation from the short wavelength (high frequency, high energy) end of the electromagnetic spectrum (Figure 13.7). 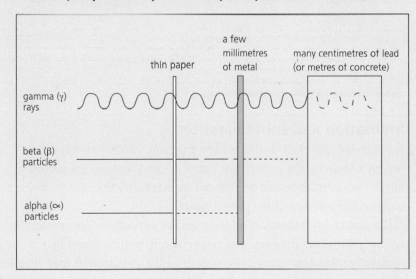 *Ref to page 158*

Two further types of 'radiation' emitted by radioactive substances are not electromagnetic waves at all but are, in fact, fast moving particles of matter which have a mass:

- beta (β) particles are electrons, so they have a negative electrical charge and only a very small mass;

- alpha (α) particles have a positive electrical charge and are several thousand times more massive than beta particles. (They are, in fact, helium nuclei, that is they consist of 2 protons and 2 neutrons.)

Because they have very short wavelengths and no charge or mass, gamma rays can pass through most materials very easily. In other words gamma radiation is very penetrating and can only be significantly reduced by sheets of dense metal (for example, lead) many centimetres thick or by concrete several metres thick.

Beta particles pass reasonably well through many materials, being gradually absorbed in proportion to the thickness of the material. They are, however, more or less completely absorbed by a sheet of any metal just a few millimetres thick.

Ref to page 158

Questions

6 Which type of radiation will cause:

a the least ionisation;

b the most ionisation;

as it passes through a substance?

7 Radioactive substances can be used to monitor the thickness of paper or metal foil as it is being manufactured.

Why are beta emitters more suitable for this purpose than alpha emitters or gamma emitters ?

Alpha particles, being larger and more massive than beta particles and carrying a greater charge are much more easily absorbed. They are more or less completely absorbed by just a few millimetres of air or by a thin sheet of paper (Figure 13.8).

The ionising effect of radiation is caused by the transfer of energy to molecules of the absorbing material, as the radiation is absorbed. The more readily radiation is absorbed, therefore, the more its ionising effect is concentrated in a short distance.

thin paper

a few millimetres of metal

many centimetres of lead (or metres of concrete)

gamma (γ) rays

beta (β) particles

alpha (∝) particles

Figure 13.8

How different types of radiation are absorbed.

cells in the testes and ovaries, for example, are more strongly affected than the cells in the hard outer parts of bones. When this has also been taken into account the resulting figure is then called the *effective dose*, also measured in sievert. Whenever you see a figure giving 'the *dose* of radiation' it is the *effective dose* that is being referred to.

Effective dose is an important quantity because it is related directly to the risk of dying from cancer. This risk can, of course, only be a best estimate in the light of all the statistical knowledge that is

Questions

8 The effective dose weighting factors for bone marrow and skin are 0.12 and 0.1 respectively.

What does this tell you about the effect of ionising radiation on these tissues?

9 What is the annual risk of death:

a for a radiation dose of 1 mSv?

b for the average UK radiation dose of 2.5 mSv?

c How does the latter compare with the annual risk of death from:

i smoking 10 cigarettes a day (1 in 200)?

ii a road accident (1 in 8000)?

10 Compile a table which briefly describes and gives the units (in both words and symbols) of: absorbed dose, equivalent dose and effective dose.

11 Which type of radiation is the most dangerous when it falls on a person's body from an external source but the least dangerous when the source is inside the cells of a person's body?

Explain your answer.

available. At the time of writing (March 2000) the International Commission on Radiation Protection estimates the probability of dying from cancer to be 0.05 Sv^{-1}. In other words the annual risk of dying from an annual radiation dose of 1 Sv is 5%, so there is a 1 in 20 chance. For a 1 mSv dose, the probability and the annual risk of death would both be a thousand times smaller. *Ref to pages 44–46*

This method of calculating risk assumes that there is a proportional relationship between effective dose and cancer risk and that all radiation doses, however small, carry some risk, that is, that there is no threshold level of dose needed to cause cancer (Figure 13.9). Although the proportional relationship is well established for moderate doses, the absence of a threshold is only assumed and is not accepted by all scientists.

Ref to Chapter 14

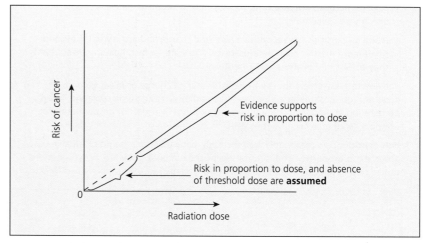

Figure 13.9

The relationship between radiation dose and cancer risk.

Irradiation and contamination

A source of radiation that is outside a person's body can harm that person's body by the radiation it emits: it can *irradiate* the person's body. This irradiation can be reduced by screening the source or moving the source and the person apart.

The source of radiation may, however, be actually on, or inside, the person's body. We then say that the person is *contaminated* by the radioactive substance that is the source of the radiation. It may then be difficult to remove the source and so that it will continue to add to the person's radiation dose for as long as it continues to emit radiation (Figure 13.10).

The ideas of irradiation and contamination can also be applied to the environment. If a source of radiation is kept securely contained, it can be used to irradiate things without contaminating them or their environment. There is then less chance of people being accidentally irradiated and no risk at all of their becoming contaminated. Furthermore, once a radioactive substance is no longer securely contained and is allowed to contaminate its immediate surroundings, it may then

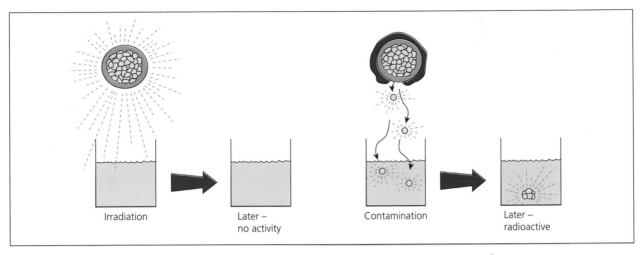

Irradiation   Later –   Contamination   Later –
               no activity   radioactive

Figure 13.10

Irradiation and contamination.

The cloud of radioactivity released by the Russian nuclear disaster at Chernobyl hit Britain yesterday.

The Sun, 3 May 1986

France said it was ready to help treat those contaminated by radiation.

The Times, 30 April 1986

...northern Europe was affected by the cloud of radiation that drifted away from the Chernobyl power station.

Today, 30 April 1986

Figure 13.11

Journalists and TV reporters often use the terms 'radiation', 'radioactive substance' and 'radioactivity' incorrectly. This may be because they do not understand the difference between irradiation and contamination.

Question

12 Re-write the extracts from newspapers shown in Figure 13.11 using the correct terms.

spread out over a very wide area as these newspaper extracts (Figure 13.11) about the Chernobyl accident show.

How do radioactive substances get moved around the environment?

Radioactive materials that are allowed to escape into the immediate environment may then be transported elsewhere by many different processses. The radioactive substances which escaped from the accident at the Chernobyl nuclear power station, for example, were carried long distances by winds and contaminated much of Europe (Figure 13.12).

Similarly, if soluble radioactive substances are dumped on landfill sites they can be carried by rainwater into the soil on agricultural land (Figure 13.13) or into drinking water supplies.

Contamination is especially serious if the radionuclide involved emits radiation at a high rate and/or if it has a long half-life.

Where does our radiation dose mainly come from?

For most people, the largest single source of the radiation that they are exposed to is the air that they breathe in their own homes, due to a gas called radon. The average radiation dose in the UK from radon is estimated as being 1.3 mSv per annum; this is quite close to the estimated world average. *Ref to pages 181-183*

Key term

Half-life. Different radioactive substances decay at different rates. The time it takes for half of the unstable atoms in a sample of a radioactive substance to decay is known as the half-life of that substance. The half-lives of different substances vary from a tiny fraction of a second to billions of years. Though there is a 50% chance of each type of unstable atom decaying one half-life, it is not possible to predict if, or when, a particular atom will decay

Figure 13.12

How the Chernobyl fall-out spread across Europe.

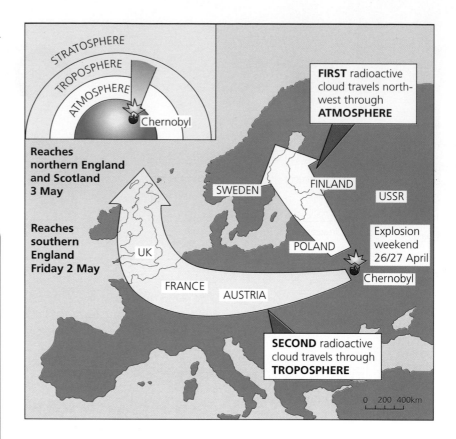

Questions

13 During the nuclear accident at Windscale (now Sellafield) in 1957, iodine-131 escaped from a chimney and fell on to surrounding fields. Cows ate the contaminated grass so radioactive iodine entered their bodies. Some of this iodine ended up in milk. Some of the infants who drank the milk received radiation doses of up to 160 mSv.

a Draw a flow diagram similar to Figure 13.13 to show how the radioactive iodine reached the affected infants.

b People's bodies retain as much of the iodine that is taken in as they need and excrete the rest.

To protect people against radioactive iodine, they may be given tablets which contain much more iodine than their bodies need.

How would this protection work?

14 Smoke detectors contain a very weak source of americium-241, an alpha emitter.

a Why are smoke detectors unlikely to be a health risk during normal use?

b Why might they be a health risk when they are eventually end up in rubbish dumps?

15 The faulty cells in hyperactive thyroid glands are sometimes killed with the radio-isotope iodine-131. This is taken by mouth and the absorbed from the bloodstream by the thyroid gland. Explain why the 8-day half-life of iodine-131 makes it suitable for this job.

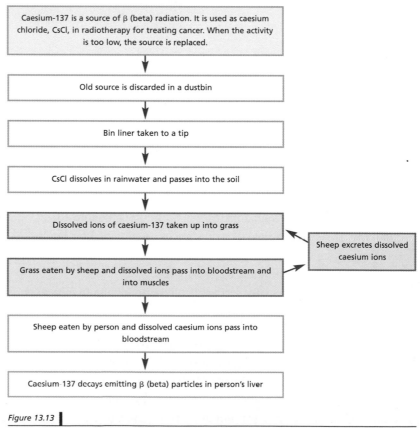

Figure 13.13

Flow diagram for the movement of a radioactive element through the environment.

Radon is a radioactive gas that is produced from a series of radioactive decays that begin with unstable uranium and thorium atoms. Since there are small quantities of these elements in almost all types of rock and soil, radon is being produced in the ground almost everywhere. Certain rocks and soils, however, contain far more than average amounts of uranium and/or thorium and so release radon at a rate that is much faster than average.

Radon-222, the isotope of radon that eventually results from the decay of uranium-238 atoms, has a half-life of only 3.8 days so that out of doors it quickly disperses and decays and isn't a significant health hazard. Indoors, however, radon levels can build up. The highest levels of radon inside buildings are found in areas where the underlying rock is limestone and at the edges of masses of granite. The levels of radon, and the resulting radiation dose, in some buildings can be hundreds of times, and sometimes thousands of times, higher than normal (see Figure 13.14).

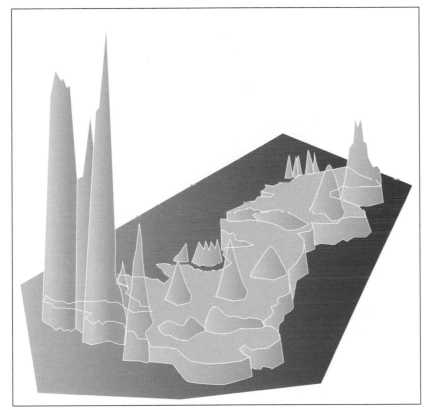

Most of the radon inside a building seeps up through the ground immediately below the building. It can be prevented from reaching the rooms inside a building by constantly pumping out the air from the space underneath the ground floor rooms.

The main health hazard from radon is not, in fact, from the decay of the radon itself. Radon is an intermediate stage in a long series of decays (Figure 13.15).

Figure 13.14

Radon doses in Britain are compared by the relative heights of the peaks on this map of the country.

Figure 13.15

The decay series from uranium-238 to stable lead-210, including radon-222.

The decay products from radon – the radon 'daughters' – are all solids, so when an atom of radon gas decays a speck of a solid substance is produced. Tiny drops of water vapour tend to form around these specks. These droplets may then be breathed in and the specks of radioactive material may lodge in, and contaminate, a person's lungs, thereby increasing the risk of lung cancer.

What other sources of radiation are we exposed to?

Though radon and its 'daughters' are the major source of the radiation to which most people are exposed, there are also various other sources. Some of these are natural sources (that is, they are simply there and there is little or nothing that we can do about them); others are artificial sources (that is, they are deliberately created by people). The artificial sources are not, of course, created in order to produce a health risk; the health risk is the price that has to be paid for using the radioactive sources for other, beneficial, purposes.

Figure 13.16 shows the contributions that different sources make to an average person's radiation dose in the UK.

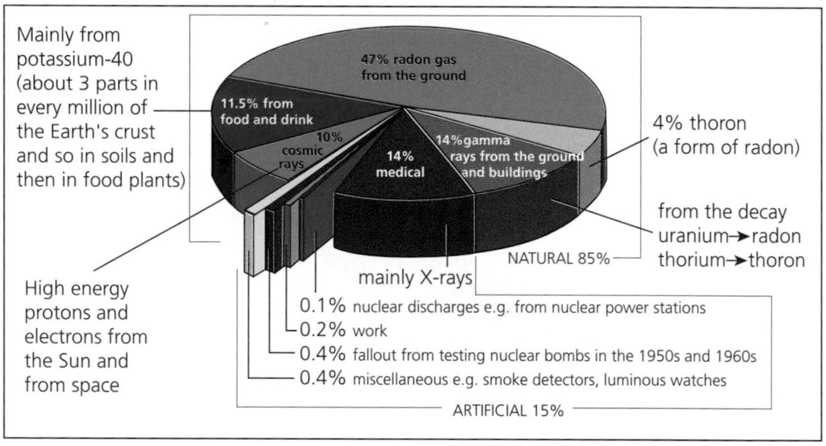

Figure 13.16

Where an average person's radiation dose comes from (UK).

Mainly from potassium-40 (about 3 parts in every million of the Earth's crust and so in soils and then in food plants)

11.5% from food and drink

10% cosmic rays

14% medical

47% radon gas from the ground

14% gamma rays from the ground and buildings

4% thoron (a form of radon)

from the decay uranium→radon thorium→thoron

NATURAL 85%

mainly X-rays

High energy protons and electrons from the Sun and from space

0.1% nuclear discharges e.g. from nuclear power stations
0.2% work
0.4% fallout from testing nuclear bombs in the 1950s and 1960s
0.4% miscellaneous e.g. smoke detectors, luminous watches

ARTIFICIAL 15%

Questions

18 About what proportion of an average person's radiation dose does radon contribute?

19 For each of the sources of radiation in Figure 13.16 say whether the danger is from irradiation or from contamination.

20 The average person's radiation dose from radioactive fallout during the mid-1960s was 0.14 mSv.

a How many times greater than the present figure is this?

b Why is the present figure much lower?

How can people be protected from radiation risks?

Since there are natural sources of radiation all around us, and even inside us, we obviously cannot be completely protected against radiation. We can, however, be protected from being unduly or unnecessarily exposed to radiation from artificial sources. Indeed various international and national bodies have been set up to ensure that the people who use sources of radiation, the people on whom these sources are used and members of the general public are all adequately protected.

The International Commission on Radiation Protection (ICRP) publishes recommendations, which are regularly updated in the light of the most recent evidence. National governments then implement these recommendations, modified where necessary to match their particular circumstances. In the UK, the National Radiological Protection Board (NRPB) advises the government about how best to implement the current ICRP recommendations.

ICRP policy is based on three fundamental principles:

1 Justification

No practice involving radiation shall be adopted unless its introduction produces a positive net benefit.

2 ALARA

All radiation exposures shall be kept As Low As Reasonably Achievable, economic and social factors being taken into account (Figure 13.17).

3 Dose limits

The radiation dose to individuals shall not exceed the limits recommended for the appropriate circumstances by the Commission.

The UK dose limits for exposure to artificial sources of radiation are, at the time of writing (March 2000):

- for radiation workers (20 mSv per annum)

- for members of the general public (1 mSv per annum)

These doses will, of course, be in addition to the average annual radiation dose, mainly from natural sources, of 2.5 mSv per annum. Also the latter figure is to be a 5-year average, thus allowing for any dosage involved in medical treatment which may occur over a short period of time. Furthermore, the recommended dose limits may be exceeded if, on balance, it is in the best interests of patients to do so, for example in the treatment of cancer.

In practice, the average additional dose for people whose work exposes them to a particular radiation risk from artificial sources is about 2 mSv, giving an overall exposure of 4.5 mSv. A small minority of workers, however, are exposed to an annual radiation dose of up to 15 mSv.

How are radiation doses monitored?

People whose work exposes them to additional radiation risks need to have their radiation dose continually monitored. A simple way of doing this is by wearing a radiation badge (Figure 13.18). The badge contains a piece of photographic film. When developed the blackening shows the amount of radiation absorbed by the film, and hence by the wearer.

Though radiation badges provide a useful indication of radiation dose, the level of irradiation of one particular part of a person's body may be very different from other parts. Furthermore, radiation badges will provide no indication of any radioactive substances internally contaminating workers.

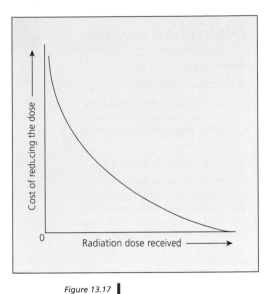

Figure 13.17

The reason for the ALARA principle. Beyond a certain point, reducing the dose further no longer justifies the cost.

Figure 13.18

Radiation detector.

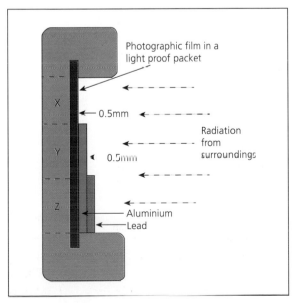

Discussion points

What are your views on:

a the ALARA principle?

b whether or not there should be different dose limits for workers and for members of the general public?

Give your reasons in each case.

Do you think your own annual radiation dose is likely to average, above average or below average? Give reasons for your answer. What additional information would you need to be more sure of your answer?

Review Questions

21 Explain:

a why a radiation badge will not detect alpha particles.

b how a radiation badge might indicate the relative levels of beta and gamma radiation.

22 Make a list of all the uses of radioactive materials that are referred to in this chapter (not forgetting the questions).

23

a Draw a 'radiation exposure profile' of a person living in the UK who is likely to receive a very much higher than normal annual radiation dose.

b Estimate what this dose might be.

c Estimate what risk this represents.

Radiation risks?

The issues

We often read in newspapers, or hear on radio and television, claims that a certain environmental factor poses a risk to health. Some of these stories concern radiation of some kind.

These claims receive a lot of media attention. If true, we might want to take some action, like moving house away from a hazard, or changing our lifestyle in some more modest way. Or we might decide that the benefits outweigh the risks. Usually there are scientists on both sides of the argument. How then can non-scientists evaluate the evidence on such claims and decide how seriously to take them?

The science behind the issues

The four case-studies in this chapter all concern radiation. Radiation is emitted by a source and travels in a straight line until it hits another object. When it is absorbed, it no longer exists as radiation but simply heats up the absorber. Ionising radiation, however, can also cause ions (charged atoms) to form, and these can cause changes within living cells, which may make them grow out of control.

What this tells us about science and society

It is often difficult to be sure whether a particular factor (for example, exposure to radiation) is correlated to a particular outcome (for example, cancer) when there are many other factors involved. One reason is that the factor may only slightly increase the risk of the outcome. The problem is made worse if the outcome is relatively rare, as there are only a few cases to consider and an apparent correlation may be simply due to chance. If several studies find the same effect, perhaps in slightly different ways, this strengthens the evidence that there is a link. People are also more likely to take a correlation seriously if there is a plausible mechanism that might explain how the factor could cause the outcome.

Because of the difficulty of getting sound evidence, scientists often disagree about whether a correlation exists. Some may change their view as new evidence becomes available. This can lead to a loss of trust in scientific knowledge, unless the media and the general public appreciate why knowledge in such cases is uncertain. Scientists who claim greater certainty than is warranted by the evidence also risk damaging this trust.

Figure 14.1

A girl with a mobile phone. Does the use of mobile phones pose a significant health risk?

Glass of wine halves chance of pregnancy

The Independent, 21 August 1998

Figure 14.2

Many science stories in the media are about claims that an environmental or dietary factor influences our health.

Toxic waste landfill sites may cause birth defects

The Scotsman, 7 August 1998

Media reports – genuine concern or just hype?

Before looking in detail at four case-studies about possible health risks due to radiation, it is worth looking at how these issues are often reported in the media (Figure 14.2).

Media reports often sensationalise new scientific claims, making them appear to be quite certain even when they are in fact provisional, based on limited evidence, not yet confirmed by other scientists, and sometimes in conflict with the published findings of other scientists. This can then lead other scientists to enter the discussion, to counter the claims being made. The result is a public disagreement between scientists. If people think of science as 'certain knowledge', this can lead them to conclude that one side or other must be either incompetent, or deliberately trying to mislead – perhaps because of whom they work for, or represent. If a scientist taking part in such a discussion is shown to have changed his or her views from the ones they expressed at some time in the past, this can also lead to suspicion.

In fact much of the uncertainty about such issues stems from the fact that good evidence is hard to obtain. Scientists cannot give a straight 'yes or no' answer, but only their best judgement at the time, in the light of the (incomplete) evidence available. This may result in considerable resentment, since quite technical matters are often involved and people realise they have no alternative but to accept the view of experts. But this expert view is often more complex and less clear-cut than people would like.

Media reports also have a tendency to give undue weight to case-histories of individual people (Figure 14.3). This may be done to add human interest to the story. But it can be very misleading, as it usually implies that there is little or no doubt that the individual's ill health was caused by whatever the 'scare' story is about.

Scientists have, however, contributed to some of these problems of communication that they now face. Where radiation is concerned,

A committee of experts has heard a huge amount of evidence from scientists and at public meetings up and down the country that masts and mobiles are ruining lives and damaging people's health.

A woman, whose husband was diagnosed with a brain tumour after a mast was put up 20 yards from their home, welcomed the news last night. 'I believe these masts are a menace and health risk,' she said.

...as time went by the family and their two children found they were not sleeping as well as they used to, an effect they now know has been linked by laboratory research to microwave radiation.

The woman then noticed that if she scratched or knocked herself, her skin was taking longer and longer to heal and her short-term memory was often poor.

Figure 14.3 Newspaper reporting of a health risk story.

there are very strong historical grounds for the media taking any evidence of the harmful effects of radiation very seriously:

> Mankind's brief acquaintance with ionising radiation seems almost to have been designed to exacerbate these feelings of ... suspicion. X-rays, radium, nuclear fission – each new discovery was greeted with wild enthusiasm, which gave way to alarm when unforeseen side-effects appeared. Protection measures were introduced, and always, sooner or later, they had to be strengthened and strengthened again.
>
> *Catherine Caufield,* Multiple Exposures, *Penguin, 1990*

Discussion point

Bearing in mind that there is, as yet, no agreement between scientists about the dangers of mobile phones:

a How fair is the first extract from the newspaper story in Figure 14.3?

b Does this family's experience provide good evidence that masts and mobiles pose a health risk?

Case-study 1: Is low-level ionising radiation a cancer risk?

Many scientists believe that there is a risk of cancer from ionising radiation, however small the radiation dose received. The main evidence is the clear correlation between radiation dose and risk, for moderate and large doses. It then seems reasonable to assume that the correlation will hold for smaller doses too. There is also a plausible mechanism: even a single alpha or beta particle, or pulse of gamma radiation, can damage the DNA in a cell in such a way as to make it cancerous. *Ref to pages 169–172*

Other scientists, however, believe that there may well be a threshold level of radiation, below which there is little or no risk of cancer. Once again, this belief isn't just a hunch: it is based on evidence, even though that evidence isn't conclusive. First, scientists agree that cells can, to some extent, repair damage that occurs to their DNA. It might, therefore, require several radiation 'hits' in a short time to cause damage to cells which they are unable to repair. Secondly, there is some evidence that cells can adapt to radiation: exposure to low radiation doses seems to make them less likely to be damaged by later doses.

If the second group of scientists are right, then many of the estimates made by the International Commission on Radiological Protection (ICRP) and the National Radiological Protection Board (NRPB), which are mainly based on low doses, would be invalid. For example, the estimated 1800 deaths per annum in the UK, mainly in Devon and Cornwall, from cancers caused by exposure to radon would be far too high (Figure 14.4).

Figure 14.4

Scientists disagree as to whether low radiation doses (for example from radon) pose a significant health risk.

Discussion points

Do you think that the ICRP and the NRPB are justified in basing their dose limits on the assumption that even small radiation doses may be harmful?

On which type of evidence would you put greater weight when deciding whether or not low-level radiation is a cancer risk: epidemiological or experimental?

Question

1 In the Columbia University study:

a Why were the results compared with cells that had not been exposed to radiation? What is the name for this second sample?

b What is the percentage increase in the incidence of cancerous mutations? Is this a convincing result? What would make it more convincing?

To try to resolve these differences and reach agreement, scientists conduct two different types of research. Firstly, they carry out laboratory experiments, usually on animals or *in vitro* (in glass) using cultures of cells. Secondly, they obtain more detailed statistical data about, for example, how the incidence of lung cancer is related to the levels of radon that people have been exposed to. These are called epidemiological studies.

 Ref to pages 50–51

How many radiation hits are needed to make a cell cancerous?

At the Centre for Radiation Research, Columbia University, New York, 250 000 mouse cells were each exposed to just a single alpha particle. Only 1 in 10 000 developed a cancerous mutation. This was almost indistinguishable from the mutation rate with no exposure to radiation at all. The experiment was then repeated using the same overall dose but applied randomly to the cells. The average radiation dose was still one alpha particle per cell, but some cells received more than an alpha particle and some received none at all. In this experiment, there were 3 cancerous mutations per 10 000 cells (Figure 14.5).

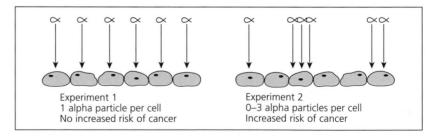

Figure 14.5

Irradiation of cells by alpha particles.

The researchers concluded that two, or more, radiation hits in quick succession were far more likely to cause cancer than an individual radiation hit. These findings seem to call into question the alleged link between exposure to radon in homes and lung cancer.

An extensive epidemiological study of radon and lung cancer

A recent epidemiological survey, however, suggests that the link between radon and lung cancer is a genuine one. Sarah Darby, at the Imperial Cancer Research Fund's Epidemiology Unit in Oxford compared the radon concentrations in the homes of 982 lung cancer victims in Devon and Cornwall with those of 3185 controls. After allowing for other variables such as smoking and social class, she found that people exposed to radon with an activity of more than 200 becquerel per cubic metre – the government's safety limit – were 20% more likely to develop lung cancer than people exposed to the average UK level of 20 becquerel per cubic metre.

A recent experimental study, by Eric Wright's team at the Medical Research Council's unit at Harwell, also lends support to the no-dose-is-safe theory. It has been found that radiation can inflict damage on a cell that only shows itself after the cell has divided several times. This seems to happen even at very low doses, for example, with just one alpha particle per cell. It has not been shown, however, that the damage to cells is the sort that makes them cancerous (Figure 14.6).

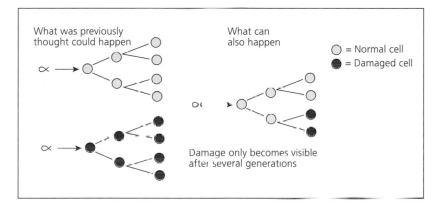

Case-study 2: The Seascale leukaemia cluster

In the West Cumbrian village of Seascale, near to the Sellafield nuclear plant, the incidence of leukaemia in children is 10 times the national average.

In 1990, Martin Gardner, an epidemiologist from the University of Southampton, claimed that he had evidence indicating that children of workers at Sellafield were between six and eight times as likely to develop leukaemia if their fathers had received cumulative radiation doses exceeding 100 mSv. A possible explanation for this was radiation damage to the fathers' sperm since it is known that reproductive cells are particularly prone to being damaged by radiation and male reproductive cells in particular are likely to receive higher radiation doses than organs deeper inside the body. The government's Health and Safety Executive, however, pointed out the puzzling fact that Gardner's findings seemed only to apply to fathers from Seascale, not to fathers who worked at Sellafield but lived elsewhere in West Cumbria.

Question

2

a What hypothesis is the Oxford study testing?

b The study compared a sample of cancer victims with a sample of non-victims. What other comparison might seem a more direct way to test the hypothesis?

c Suggest a reason why the researchers chose to compare these two groups.

Key term

The **activity** of a sample of radioactive material is measured in becquerel (Bq). 1 becquerel (Bq) is one nuclear decay per second.

Figure 14.6

Effect of alpha particles on cells.

Discussion points

Do you think the balance of evidence suggests that any radiation dose, however small, carries a risk of cancer? Or do you think that low doses have no risk? Or do you think it is impossible to be sure? What additional information would you most like to have to help you reach a decision, or to check your current view? Would it be possible to obtain this information? If so, how could it be obtained?

Key term

The **cumulative** dose means the *total* dose added up over the years – as opposed to the *average* dose per year.

Families whose children had leukaemia used Gardner's findings as evidence when claiming damages from British Nuclear Fuels Limited (BNFL). However, in 1993, Mr. Justice French, a High Court judge, ruled against the families on the grounds that there was insufficient evidence. During the 90-day hearing, BNFL had argued that the Gardner study was based on too few cases to be reliable and that his findings were inconsistent with the findings from a far larger number of cases in Japan after the nuclear bombs were dropped on Hiroshima and Nagasaki. They suggested that the cluster of childhood leukaemia cases in Seascale was due purely to chance or was possibly caused by an unidentified virus.

Figure 14.7

Children at Seascale.

Questions

3 What were BNFL's main arguments against Gardner's conclusions?

4 Which type of radiation, alpha, beta or gamma, is more likely to damage the sperm of male workers than the ova of female workers? Explain your answer.

Discussion point

Some people argue that it is ethically unacceptable to carry out experiments of this sort using animals. List the main arguments for and against. *Ref to pages 35–37*

Discussion point

Do you think the balance of evidence suggests that the childhood leukaemia cluster around Seascale is due to the Sellafield plant? Or do you think there is some other cause? Or are you not sure? What additional information would you most like to have to help you reach a decision, or to check your current view? Would it be possible to obtain this information? If so, how could it be obtained?

Some experimental research during the late 1990s, however, indicated that Gardner may, after all, have been right. A team of British researchers produced evidence that two hits by carcinogens (cancer-causing agents), one to a father and one to his offspring, can trigger leukaemia. All of the offspring of male mice that had been injected with plutonium developed either leukaemia or lymphoma (another type of cancer affecting lymph nodes) after they were exposed to a chemical carcinogen. Only 68% of mice whose fathers had not been injected with plutonium developed cancer after exposure to the same dose of the same chemical carcinogen.

A large-scale epidemiological study was also undertaken during the late 1990s in order to confirm, or disprove, the Gardner theory. This involved 40 000 children born to 18 000 workers in the nuclear industry at several sites in the UK. The report, published in 1999, concluded that the children of fathers who had been exposed to radiation before their conception were more likely to have leukaemia than the children of fathers who had not been exposed to radiation. The risk was greater for the children of fathers whose radiation dose was more than 100 mSv. The study also showed, however, that the overall rate of all types of cancer among the children was not significantly greater than in the UK population as a whole. The study failed, therefore, to settle the issue, concluding only that Gardner's work was 'not disproved'.

But then, later in 1999, there was a new twist to the story. Evidence was found supporting the 'virus' theory that had originally been suggested 10 years earlier (Figure 14.8).

Figure 14.8

New evidence on the Seascale leukaemia cluster.

The cluster culprit
Childhood Leukaemia is probably an infectious disease

A major new epidemiological study has turned up the best evidence yet that leukaemia "clusters" are triggered by infections that occur in populations where many of the people are new to the area.

The best known of Britain's childhood leukaemia clusters is in the village of Seascale in Cumbria, close to the Sellafield nuclear plant, where the disease is 10 times as common as it is in Britain overall. Radiation was initially blamed, but in 1988, Leo Kinlen of the University of Oxford suggested a "population mixing" theory. When large numbers of workers move into an isolated area, Kinlen argued, the immunity passed down by migrant mothers may not prepare their children to fight off the particular strains of viruses and bacteria found in the area.

Heather Dickinson and Louise Parker of the University of Newcastle have now scrutinised the health records of 120 000 children born in Cumbria between 1969 and 1989. They found that if both parents came from outside Cumbria, the chance that their child would develop leukaemia during the first 6 years of their life was 2.5 times as high as when one or both parents came from Cumbria. In areas where the population influx was heaviest, and up to 80 per cent of the parents were migrants, the risk was up to 11 times as high as in the areas with the fewest migrants.

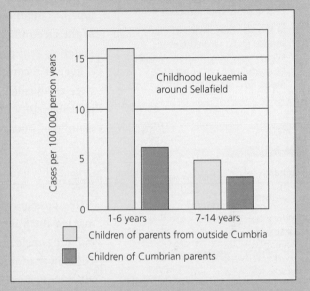

"I would say that Kinlen's infection hypothesis is now established," says Richard Doll of the University of Oxford, the epidemiologist who proved the link between smoking and lung cancer.

New Scientist, 21 August 1999

Case-study 3: Are mobile phones a health risk?

There has been a huge increase in the numbers of mobile phones in use in Britain since these were first introduced in the early 1990s. Because mobile phones are now so common, and so popular, any 'scare' story about their possible dangers is sure to be big news. The link between mobile phones and brain tumours (cancer) is not, however, nearly so well established as many media reports appear to suggest (Figure 14.3).

An epidemiological study from Sweden, published in 1999, involved 209 people who had brain tumours and over 400 matched controls. The study indicated that mobile phone users have a 2½ times greater chance of getting a brain tumour in the part of their brain close to their 'phone ear'. There was, however, no greater overall risk of a brain tumour amongst mobile phone users, so the Swedish evidence is not compelling. The total number of each type of brain tumour was quite small, so any apparent correlations could be simply due to chance.

A similar study in the US, also published in 1999, and involving more than twice as many people with brain tumours, also found no

Figure 14.9

Radiation from a mobile phone passing through the brain.

5

a What hypothesis was being tested in the Swedish study?

b Why do you think the researchers did not choose to compare the incidence of brain tumours in a sample of mobile phone users and non-users?

6 In the US study, twice as many sounds like a large increase. What other information would be useful to have in order to judge how significant it is?

overall difference in the overall number of brain tumours between users and non-users of mobile phones. There was an apparent difference, however, in the types of tumour: nearly twice as many mobile phone users had a type of tumour called a neurocytoma.

Then there is the issue of a plausible mechanism. Case-studies 1 and 2 both involved ionising radiation. Scientists *know* that this type of radiation can cause cancer and also have a pretty good understanding of *how* it causes cancer. Mobile phones, however, use microwaves. This type of radiation is towards the long wavelength (low frequency) end of the electromagnetic spectrum and does not cause ionisation in the materials that absorb it (Figure 14.10).

Even if a correlation between microwave radiation and brain cancer were established, therefore, scientists would still need to provide a plausible story of how microwaves could affect DNA before they could claim that microwaves *cause* cancer.

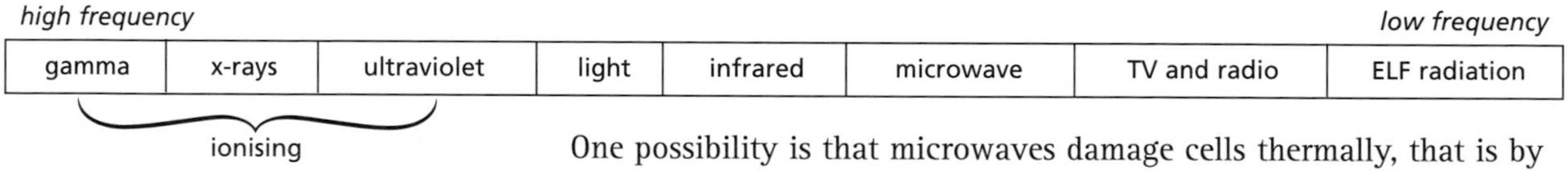

high frequency   low frequency

gamma	x-rays	ultraviolet	light	infrared	microwave	TV and radio	ELF radiation

ionising

Figure 14.10

The electromagnetic spectrum.

One possibility is that microwaves damage cells thermally, that is by making them hotter, just as food becomes hotter by absorbing the radiation inside a microwave oven. This, however, is not very likely. The power of microwave phones (the rate at which they emit energy via microwaves) is less than a thousandth of the power of an average microwave oven. Even allowing for a possible focusing effect of the brain tissue, creating 'hot spots', the microwave radiation from a mobile phone is unlikely to heat up brain cells as much as they are heated up when a person takes even mild exercise. The radiation from microwave masts is, of course, more powerful than from the phones themselves, but it is many times further from anyone's head and the intensity of the radiation falls off in proportion to the square of the distance. In other words, if the distance from the source of radiation is increased by a factor of 10, the intensity of the radiation reduces by a factor of 100.

Scientists investigating the possible link between mobile phones and brain cancer have, therefore, been looking for other evidence of how microwaves might damage cells, usually by using animals or *in vitro* cultures of cells. Various claims have been made, for example that microwaves can cause strands of DNA to break, just as ionising radiation can, or that microwaves can cause an increase in certain types of cancer in mice (Figure 14.11). Other researchers, however, have not been able to repeat these results.

There does appear to be evidence that microwaves make tiny nematode worms grow faster and this may be due to an effect on the genes that can cause cancer when they are damaged. There is also evidence that microwaves change the rate at which certain chemicals are produced in the brains of rats and that this may be related to their lack of alertness. Studies in humans, however, have shown no effect of microwaves on

Discussion point

Scientists usually like to carry out experiments which lead to new knowledge. Why might it also be important to repeat an experiment that someone has already done?

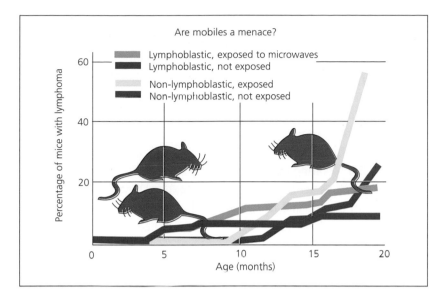

Are mobiles a menace?

- Lymphoblastic, exposed to microwaves
- Lymphoblastic, not exposed
- Non-lymphoblastic, exposed
- Non-lymphoblastic, not exposed

Percentage of mice with lymphoma (y-axis: 0, 20, 40, 60)
Age (months) (x-axis: 0, 5, 10, 15, 20)

Figure 14.11

Incidence of lymphoma in mice exposed to microwave radiation (from New Scientist, 10 May 1997).

Questions

7 Explain why the intensity of radiation falls off as you go further from the source. Can you suggest why it falls to $1/4$ if you go twice as far away?

8 What does Figure 14.11 suggest about the carcinogenic effects of microwave radiation on mice?

Discussion point

Do you think the balance of evidence suggests that mobile phones can lead to brain tumours? Or do you think that the evidence shows that mobile phones are safe? Or do you think it is impossible to be sure? What additional information would you most like to have to help you reach a decision, or to check your current view? Would it be possible to obtain this information? If so, how could it be obtained?

short-term memory and a *reduced* reaction time which indicates *greater* alertness. Other studies have shown that cells in the hippocampus region of rats' brains are especially affected by microwaves but the hippocampus is deep inside the human brain and is unlikely to be affected by very weak microwave radiation. To date, no one has come up with a story that a majority of other scientists find convincing about how low-level microwave radiation could cause brain tumours.

Case-study 4: Living close to a high-voltage power line

Pylon lines are a feature of our landscape, both in cities and in the country. They carry electricity from power stations to consumers. We may regard them as rather ugly. Pylons carry safety warnings about the dangers of high voltage cables, but until around 1980 no one had suggested there was any other danger from power lines. Then a study in the United States suggested a higher incidence of cases of childhood cancer among children living near a power line (Figure 14.12).

We don't normally think of cables and power lines carrying mains electricity as emitting radiation. But mains electricity in the UK is an alternating 50 hertz (Hz) supply, that is, the current changes direction and back again 50 times each second. This means that electrical and magnetic fields around cables and power lines

Figure 14.12

Power lines over a housing estate in Manchester.

are also alternating with the same frequency. As a result, the cables emit electromagnetic radiation. Ref to page 158

The radiation from wires and cables carrying mains electricity has a very low frequency compared to other types of radiation in the electro-magnetic spectrum and is sometimes referred to as extra low frequency (ELF) radiation (Figure 14.10).

ELF radiation, like microwave radiation, is non-ionising. In fact, its only known effect is that it can, like microwaves and radio/TV waves, cause small alternating currents to flow in materials that conduct electricity, such as aerials and, to a lesser extent, living cells. It also sets up small electric and magnetic fields within materials. The size of this magnetic field decreases rapidly with distance from the power line (Figure 14.13).

Ordinary household electrical appliances, however, also emit ELF radiation and generate small magnetic fields around them when they are switched on. As Figure 14.14 shows, these can be larger than the field 25 metres from a power line. On the other hand, many appliances are only switched on for short periods of time, whereas fields due to power lines are always present. As there have never been suggestions that domestic electrical appliances pose a cancer risk, scientists were inclined to dismiss the claims about power lines.

It was a Swedish study reported in 1993 that caused the scientific community, and the bodies concerned with health and safety regulations, to take more seriously

Figure 14.13

Magnetic field strength under a power line.

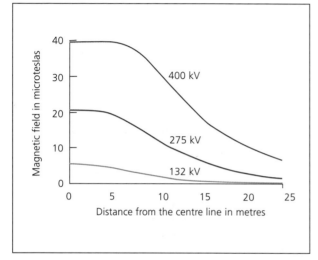

Key term

The **microtesla** is a unit of magnetic field strength.

Figure 14.14 Exposure to magnetic fields (from New Scientist, 7 October 1994).

Household appliances (30 cm away)	Magnetic field strength/microteslas
Vacuum cleaners and drills	2–20
Food mixers	0.6–10
Hair driers	0.01–7
Dishwashers	0.6–3
Washing machines	0.15–3
Fluorescent lamps	0.5–2
Electric ovens	0.15–0.5
Beneath power lines	
Immediately beneath 400 kV line	40
25 metres away	8
Immediately beneath 275 kV line	22
25 metres away	4
Immediately beneath 132 kV line	7
25 metres away	0.5

the suggestion of a link between power lines and health. The researchers identified all 127 383 children in Sweden (under age 16) who had lived within 300 metres of a power line between 1960 and 1985. In this sample, they found 141 cases of cancer, including 39 of leukaemia. They looked for four controls for each case (children who were the same in all respects except that they had not lived near a power line). They took great care over this matching, looking for children from the same area, of the same sex and age. They eventually identified 554. Using records from the electricity companies they then worked out the exposure of each child to ELF radiation and magnetic fields.

In one analysis of leukaemia cases, they compared children exposed to average magnetic fields over 0.2 microtesla with those exposed to average fields below 0.1 microtesla (Figure 14.15). Children exposed to the higher field were 2.7 times more likely to develop leukaemia. This rose to 3.8 times for children exposed to a magnetic field of average strength 0.3 microtesla.

Because of the care with which the samples were chosen and the information collected, and the size of the study, these findings were generally felt to be more reliable than many epidemiological studies. It was unlikely that other environmental factors were responsible for the results. On the other hand, the numbers of people contracting cancers was small. The researchers estimated that power lines account for 3.5 cancer cases per annum in the whole of Sweden.

To be able to claim a causal connection, however, scientists need to be able to explain how ELF radiation might cause cancer. Some research published in 1995 seemed to provide the necessary explanation. When cell cultures were subjected to ELF radiation, cells were found to divide more rapidly and to produce abnormally large amounts of messenger RNA, molecules which are the intermediate step between genes (DNA) and proteins. Since cancer is caused by uncontrolled cell division, this was the required link with leukaemia. Unfortunately, however, other researchers have been unable to replicate these results. It has been suggested that contamination of glassware by tiny particles of a magnetic substance called magnetite – which is known to happen quite often in laboratories – could have been responsible for the original findings. There is, as yet, no conclusive evidence that ELF radiation can cause changes to a cell's DNA.

Denis Henshaw and his colleagues at the University of Bristol suggested, in 1996, that there may be a link between strong electrical fields and radon. The fields may raise the concentration locally of tiny water droplets containing radon, making it more likely that people in the vicinity will inhale them. These conclusions are not, however, generally accepted.

The case remains open. In 1998, a panel of experts in the US classified strong electrical and magnetic fields as 'possible human carcinogens', the lowest rating for a suspected cancer hazard and urged further experimental and epidemiological research.

Questions

9

a How does the magnetic field due to a high voltage power line 25 metres from a house compare with that due to the appliances inside a house?

b On the basis of Figs. 14.13 and 14.14, how close to power lines do you think houses should be built?

10 From Figure 14.15 calculate the ratio of cases to controls for the higher exposure group, and for the lower exposure group. How much bigger is this ratio for the higher exposure group?

11 What features of the Swedish study made scientists take its findings seriously?

Figure 14.15 Cases of leukaemia among Swedish children under 15 exposed to larger and smaller magnetic fields

	Cases	Controls
Exposed to larger fields (>0.2 microtesla)	7	46
Exposed to smaller fields (<0.1 microtesla)	27	475

Discussion point

Do you think the balance of evidence suggests that living near high voltage power lines increases the risk of cancer? Or do you think that power lines pose no risk? Or do you think it is impossible to be sure? What additional information would you most like to have to help you reach a decision, or to check your current view? Would it be possible to obtain this information? If so, how could it be obtained?

Thinking about radiation risks

What, then, can we learn from the four case studies summarised in this chapter? One conclusion is that we need to know more about all of them to reach an informed decision. The accounts above are very brief and really little more than outlines. If you want to know more, then you will find books and articles on all these issues, and information on the worldwide web.

The cases all show how difficult it is to obtain conclusive evidence that a factor causes an effect – or that it doesn't. Some of the reasons for this are outlined on pages 47–48. Have a look at that section and try to relate the general issues it raises to each of these four cases.

Ref to pages 47–48

One obvious problem is that all of these claimed health risks are rather low. Of course, in a practical sense, this is good news – but it makes it hard to establish if there is any risk at all. Most outcomes are the result of a combination of factors, both environmental and genetic ones. So it can be hard to spot a factor that might be worth investigating more deeply, as so many things vary from person to person. Once a factor is suspected of having an effect, epidemiological studies can help to provide evidence of an association. Getting convincing evidence is very hard if the number of cases is small (that is, if the incidence of the condition is low). If only small numbers are involved, any difference may be due to random variation and not a real effect at all. Large samples are often needed to sort this out – and studies with large numbers take a lot of time and cost a lot of money.

Finding an association (or correlation) is not the same as establishing a cause. The two things might both be caused by another factor. Scientists are more likely to take a correlation seriously if there is a plausible mechanism which could link the two. Sometimes this evidence comes from studies in the laboratory, for example, using cell cultures. Sometimes it comes from research on animals, raising ethical issues about the use of animals – and also problems of interpretation (how far do the findings also apply to humans).

All the uncertainties surrounding issues like these can cause problems for science communication. Controversial issues are often given a high profile in the media and are regarded by many members of the general public as being very important. It is not surprising, therefore, that people can be upset to find that scientists themselves are unsure about whether or not radiation of various types and/or levels is a health risk. Journalists may pick up some rather provisional findings and report these more prominently than they warrant. We need, as a society, to improve the ways we report and discuss such issues, to build rather than damage trust between scientists and the public.

Discussion points

Explain, with examples, the difference between experimental and epidemiological evidence for the link between radiation and cancer.

When mobile phone companies announced that they were developing phones which used lower power microwaves and which shielded users against them, they were accused of admitting that microwave radiation was harmful.

a Was this accusation fair?

b What other reasons might phone companies have had for spending money on these developments?

12 Read the following passage. Then answer the questions which follow.

'Unable to understand the technology that affects much of our lives, we are increasingly reliant on various specialists ... to interpret these technologies for us. ...

Unfortunately, scientists have all too often tried to gain the public's confidence by making misleading statements. Part of the trouble is that scientists think differently from the rest of us. To a scientist there is no absolute certainty, no immutable fact – only probabilities and an evolving understanding of how the world works. A simple statement such as "this X-ray is perfectly safe" ... may mean very different things to the layperson and the scientist. A scientist will recognise the unspoken provisos in the statement and know that a more reasonable rendering would be "as far as we know, this X-ray has a very low probability of causing you detectable harm". ... To a layman, of course, the original statement sounds very much like a guarantee or a promise. When, as often happens, these "promises" are broken, the public begins to develop a distrust of science and scientists – a distrust that may grow to irrational proportions.'

Catherine Caufield, Multiple Exposures, *Penguin, 1990*

a How far does the information in this chapter support Caufield's point of view? What other reasons could there be for the public's disenchantment with scientists?

b How widespread do you think the public distrust of science and scientists actually is?

c What could be done to improve the situation:

i by scientists themselves?

ii by members of the general public?

iii by the education system?

Understanding the solar system

The issues

Have you ever looked up at the sky and wondered what the objects are that you can see, and how far away they are? If so, you are in good company. From the earliest recorded history, people have been asking questions about the universe. The Sun, Moon and stars are a source of wonder as well as curiosity – they have inspired stories and poetry. Curiosity about the universe is an important expression of what it means to be human. Our ideas about the universe are important because they profoundly affect the way we see ourselves, and our understanding of who we are.

Figure 15.1

Stonehenge is like an ancient observatory. The stones are aligned with the rising sun on midsummer and midwinter day. Other stones are aligned with certain positions of the moon.

The science behind the issues

Commonsense tells us that we are living on a large and fairly flat surface, which is fixed and stationary. Science tells a different story: we live on the surface of a sphere which orbits the Sun (once a year), while also spinning on its axis (once a day). The Sun is a star, and the Earth is one of nine planets which make up the solar system.

Figure 15.2

The domes of the Keck observatories which house optical telescopes high up on Mauna Kea in Hawaii. Each telescope has a large mirror with a diameter of 10 metres made up of 36 smaller hexagonal mirrors.

What this tells us about science and society

Understanding the motion of the stars and planets required accurate observation and imagination. All we have to go on are observations made from Earth (plus, more recently, some observations made from satellites in Earth orbit and from space probes). New instruments, like the telescope, were vital for making progress. An exact match between data and prediction was important in convincing people. A mechanism which accounted for the motions (Newton's idea of universal gravitation) helped to consolidate this. When the model was used to predict a new planet, and this was then found, it increased confidence further. Sending spacecraft close to the outer planets is a dramatic demonstration of the accuracy of our model of the solar system.

The scientific model of the solar system

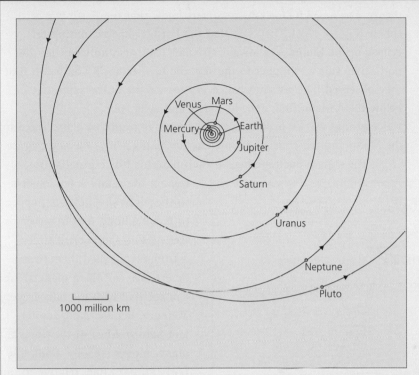

Figure 15.3

The orbits of the planets, drawn to scale.

Some other facts about the solar system

- All the planets rotate as they orbit. This causes day and night on the planet.

- The Earth's axis of rotation is tilted relative to the plane of its orbit. This causes the seasons. In the summer, the northern hemisphere points towards the Sun, so days are longer and the average temperature is higher.

- The orbits of the planets are nearly (though not exactly) in the same plane.

- There is also a large group of tiny planets (the asteroids), orbiting mainly in the gap between Mars and Jupiter.

- A large number of comets also orbit the Sun. These move in very elongated elliptical orbits, so most of the time they are too far away from the Sun for us to see them. Their orbits are not close to the plane of the planets' orbits.

Figure 15.4

The nine planets, drawn to scale.

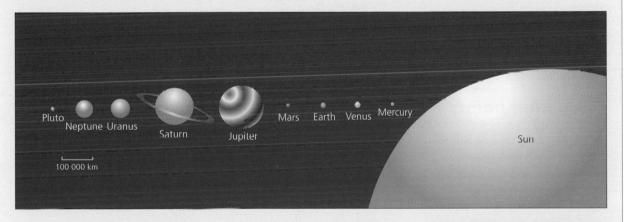

Figure 15.5 Data on the solar system.

	Diameter (km)	Mean distance from Sun (million km)	Time for one orbit
Sun	1 392 000	-	-
Mercury	4 878	57.9	88 days
Venus	12 103	108.2	225 days
Earth	12 756	149.6	365 days
Mars	6 786	227.9	687 days
Jupiter	142 980	778.3	11.9 years
Saturn	120 540	1 427	29.5 years
Uranus	51 120	2 869.6	84 years
Neptune	49 530	4 496.6	165 years
Pluto	2 280	5 900	248 years

Questions

1 The distance from the Earth to the Sun is sometimes called 1 astronomical unit (1 AU). What is the distance from the Sun to Neptune, in AU?

2 In a scale model of the solar system, the Sun is a large ball of diameter 50 cm. What size should the Earth be, and how far should it be from the Sun?

Ancient astronomy

The idea that the Earth is flat has never been an accepted scientific theory. The ancient Greek astronomers knew that the Earth was a sphere (Figure 15.6). They saw that the Earth's shadow, during an eclipse of the Moon, was always circular. They also noticed that the north pole star was higher in the sky, the further north you went. And they observed that, as ships sailed out to sea, the hull disappeared below the horizon first, and the masts and sails later.

Everything we see in the sky seems to move round us daily. The Sun, the Moon and the stars all follow circular paths across the sky (Figure 15.7). The stars move together, so their patterns (the constellations) remain the same. A few objects in the night sky, however, appear to move among the stars. For this reason, the ancient Greek astronomers called them 'wanderers'. We call them planets. We can see five with the naked eye: Mercury, Venus, Mars, Jupiter and Saturn. Most of the time, a planet moves steadily along a fairly straight line, relative to the background of stars. But from time to time, it appears to move backwards, and retrace its steps (retrograde motion) (Figure 15.8).

All of these observations were known to the ancient Greek astronomers. Around 500 BC, the followers of Pythagoras suggested that the Earth was at the centre of a set of crystalline spheres which moved round, carrying the stars, Sun, Moon and planets with them. This explains the basic motions we observe, but doesn't account for the details, especially the retrograde motion of the planets. Around 370 BC, Eudoxus came up with a way of explaining these (Figure 15.9).

Stick set up at Alexandria

Parallel rays of sunlight

Shadow cast by stick

a

y

S

Stick set up at Syrene (no shadow)

Discussion points

What is astronomy? How does it differ from astrology?

What evidence would you use to convince someone that the Earth was a sphere?

Figure 15.7

A long-exposure photograph of the night sky. The stars appear to move in circular paths, around the north pole star.

Aristotle, who lived from 384 to 322 BC, was the most famous philosopher of his day – and one of the most influential ever. Starting from the ideas of Euxodus, he developed a model with 55 spheres. The Earth was

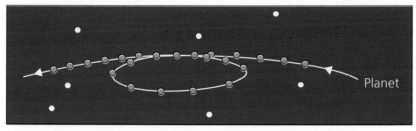

fixed and at the centre. All the heavenly bodies were carried round on spheres, as this was regarded as the 'perfect' shape. The heavens, according to Aristotle, are perfect and unchanging – made of a different kind of matter from that on Earth. On Earth, there is imperfection and change.

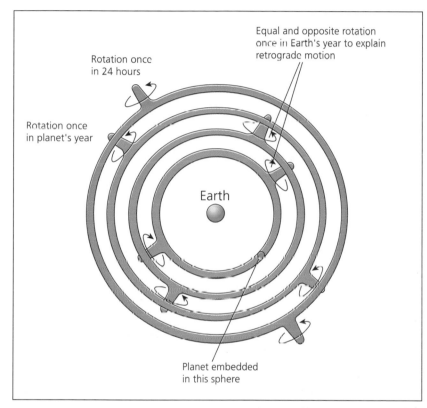

Rotation once in 24 hours

Equal and opposite rotation once in Earth's year to explain retrograde motion

Rotation once in planet's year

Earth

Planet embedded in this sphere

> ## Key term
>
> Aristotle's model is a **geocentric** one – the Earth is at the centre.

Figure 15.9

Eudoxus's model has sets of spheres, mounted one inside the other. Each rotates independently, and together this produces the motions we observe.

Figure 15.10

In Ptolemy's model, each planet moves round a small circle, whose centre circles the earth. The combination of these two motions produces the effects we observe – including retrograde motion.

Even with 55 spheres, this model could not predict the exact positions of the planets. Then, around 230 BC, Apollonius of Perga proposed another idea. He suggested that each planet moves in a small circle, whose centre moves round a larger circle with the Earth at its centre (Figure 15.10). The effect is a bit like a fairground ride with cabins that spin on their own axes, while being carried round on a larger rotating frame.

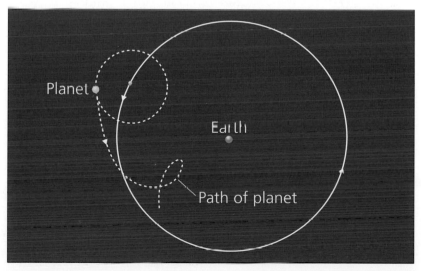

Planet

Earth

Path of planet

Question

3 Use Figure 15.10 to explain why a planet sometimes appears to move backwards against the stars.

Discussion point

Ptolemy's model was not challenged for over 1200 years. Why do you think it survived so long?

Key term

A **heliocentric** model is one which puts the sun at the centre.

This idea was picked up and developed by Claudius Ptolemy, who lived from about AD 100-170. He made some modifications (like having some circles a little off-centre, and allowing the planet's speed to change as it moved round) and added some extra circles. His most famous book, the *Almagest*, described this very complex model. It had become so complex that most people no longer believed that these spheres and circles really existed – they were just a way of calculating the positions of the planets. It was said to 'save the appearances'. But it worked – and remarkably well. It predicted the positions of the planets so well that nothing better was produced for over 1000 years. Although we now have a better way to look at these motions, Ptolemy's model is a remarkable achievement.

The Copernican revolution

The ideas of Aristotle, and Ptolemy's model of the universe, were widely accepted and became very influential. Around AD 1260, Thomas Aquinas, a Christian monk, developed a philosophy which combined Aristotle's ideas with Christian teaching. This became the accepted viewpoint of the Church. The heavens were the realm of God, and hence perfect. And it was God who kept the crystal spheres moving.

However, astronomers were becoming increasingly aware that Ptolemy's model did not lead to completely accurate predictions. Also it seemed impossible to *explain* such complex motions: what could make things move in this way? Early in the 16th century, Nicolaus Copernicus (a Polish mathematician and astronomer who was also a canon in the Church) wrote a book which started a revolution in thinking about the universe. In it he suggested that the motions of the stars and planets would look much simpler if the Sun is put at the centre, with the Earth and the other planets circling around it (a *heliocentric* model) (Figure 15.11). The Earth is a planet. It also spins on its axis, which causes day and night, and explains why we see all the celestial bodies moving across the sky (Figure 15.7).

This can explain the retrograde motion of the planets (Figure 15.12). As the Earth overtakes a planet further from the Sun (like Mars), the planet seems to move backwards against the background of the more distant stars.

Figure 15.11

Copernicus's model of the solar system. This is a simplified version of the diagram in his book De Revolutionibus.

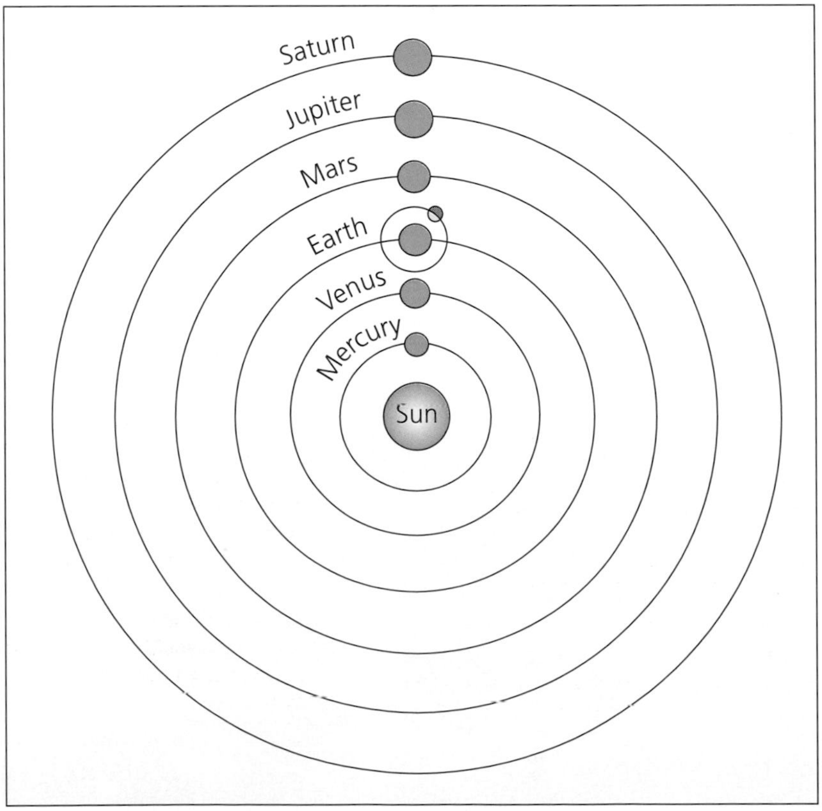

Copernicus thought long and hard before publishing his ideas. He started to write his book (*De Revolutionibus Orbium Coelestium*, or *On the Revolutions of the Heavenly Spheres*) in 1509, but it was only published shortly after his death in 1543. His main worry was that his ideas would be ridiculed by other astronomers. So he spent many years checking his model against observation, and improving it. In the process he made it much more complicated than Figure 15.11. In fact he added so many extra circles and other modifications that it ended up almost as complicated as Ptolemy's.

Copernicus may also have been worried about opposition from the Church. When his book was eventually published it contained a preface written by Andreas Osiander, a Lutheran cleric, which explained that the model was just a means of calculating the planets' positions more simply and not a description of how things really were. This, however, does not seem to have been Copernicus's own view.

From our perspective today, it is easy to think of Copernicus's ideas as 'obviously' correct. At the time, however, it was far from obvious. There were some persuasive arguments in favour of a geocentric model.

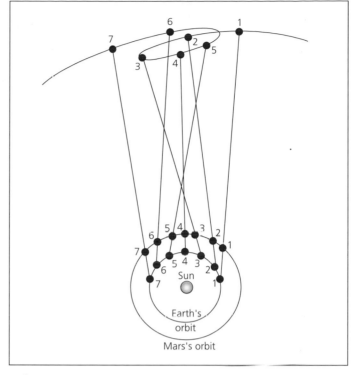

Figure 15.12

The retrograde motion of planets is due to the fact that they orbit at different speeds. So as the Earth overtakes another planet, its position against the background of stars changes, and appears to loop backwards.

The Geocentric Model

Arguments in favour of the geocentric model ...	How a Copernican might respond ...
The Earth *feels* as though it is fixed, not flying through space and rotating on its axis.	We are moving along with it. If you are in a moving vehicle, you often cannot tell that you are moving.
If the Earth rotates once per day, this means that the whole surface is moving at a high speed. So why do we not feel a strong wind, as the Earth's surface rushes along?	The Earth's atmosphere moves along with it – so we do not experience a wind.
If we throw something straight up, it comes back to our hand. Surely if we were rotating at high speed, it would land a long way away, because we would have moved while it was in the air.	This is a common misconception. If you try this (for instance, by throwing a ball upwards while you are running along), the ball comes back to your hand. A projectile has the same forward motion as the object that released it.
If we observe a star now, and then again in 6 months' time (when the Earth is at the other side of its orbit), it should be in a different direction. (This is called *parallax*.) But in fact we do not observe this.	This is because the stars are very very far away. The change is too small to observe. (In fact, it was first observed in 1838.)

⟳ Ref to page 209

Ref to page 209

Question

4 What were Copernicus's two main worries about the reception of his ideas?

  Why do you think some people in the Church might have opposed the heliocentric model?

Discussion points

Are you more persuaded by the arguments and evidence *for* or *against* the heliocentric model? What do you consider the strongest argument for the heliocentric model? What is the strongest argument against it?

If you had been alive in 1543, do you think you would have accepted the heliocentric model or the geocentric one?

Discussion points

Both Copernicus's model and Ptolemy's model lead to prdictions which are roughly equally accurate. Does this mean that there is no way to decide which is the better model?

Do you think it is possible to *deduce* the heliocentric model from the data? If so, why did no one do it before Copernicus? If not, how did Copernicus arrive at it?

Copernicus's book did not cause an immediate stir. In fact, few people appeared to notice it, perhaps because it was complicated and detailed, and seemed quite similar to many older books on the Ptolemaic system. It was not immediately banned by the Church. But in the years that followed, social and political events led to a hardening of attitudes, and a greater fear of unorthodox ideas. One major factor was the Counter Reformation – the response of the Roman Catholic Church to the Reformation. Between 1545 and 1563, a great Council met at Trent in northern Italy. It reaffirmed traditional doctrine, founded the Jesuit order to oppose the reformers, and made a list (or Index) of forbidden books, which the Inquisition would enforce. This was to have an important influence on the events which followed.

Convincing others

The heliocentric model of the solar system, proposed by Copernicus, was not immediately adopted by astronomers. There was some opposition from scholars steeped in the Aristotelian view, and also from some in the Church. And, as we have seen, the evidence for the heliocentric model was not overwhelming. Three people played key roles in getting the heliocentric model generally accepted: Tycho Brahe, Johannes Kepler and Galileo Galilei.

Tycho Brahe: the great observer

In 1559 the 13-year-old Tycho Brahe decided to become an astronomer when he observed a partial solar eclipse. What impressed him was not the eclipse itself but the fact that astronomers had been able to predict it. Tycho, the son of a wealthy Danish family, was able to pursue his interest in astronomy. On the night of 24 August 1563, something happened that was to shape the rest of his career. There was a conjunction of Jupiter and Saturn – that is, the two planets appeared so close together in the sky as to seem like one bright object. Like the eclipse, this had been predicted – but the prediction was several days out.

This was not good enough for Tycho. It convinced him that Ptolemy's model had to be improved. And for that, we would need more accurate data. By the age of 26 he had set up the greatest observatory in Europe, with instruments provided by the Danish king. Chief among these was a huge quadrant (Figure 15.13), with which he could measure the positions of stars and planets to an

Figure 15.13

At a time when most astronomers were using sticks and string to measure the position of stars, Tycho built a quadrant (like a large protractor) over 2.5 metres high. This enabled him to make very accurate observations.

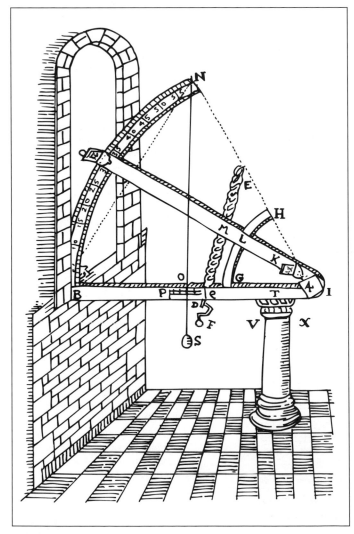

accuracy of one minute of arc ($1/60°$). Over 20 years, Tycho built up a huge and very accurate record of the positions of all the planets.

Tycho himself could not accept the heliocentric model; he seems to have thought that the idea of the Earth moving was simply inconceivable. He also knew that the stars showed no parallax shift – even with his excellent instruments (Figure 15.13). But two particular observations led him to reject the Ptolemaic model of crystalline spheres. The first came in 1572, when Tycho observed a supernova – a very bright new star which appeared for a few months then gradually dimmed. This challenged the accepted view that the heavens were eternal and unchanging. Then in 1577, he observed a comet and tried to estimate its distance by making parallax measurements from two places on the Earth's surface. By studying it over several weeks, he concluded that its path took it through the orbits of several planets. This meant that the planets could not be mounted on crystalline spheres. Rather than switch to the Copernican model, though, Tycho proposed a compromise: the Earth at the centre, the Moon and Sun going round it, and the other planets going round the Sun. Tycho's model never really caught on. But the data he had collected provided the springboard for Johannes Kepler to make the decisive breakthrough.

Johannes Kepler: a model that fits the data

Johannes Kepler was born in 1571 in Weil der Stadt near Stuttgart. Although he came from a 'problem family' and was a rather solitary child with few friends, his abilities were recognised at school and he went on to the local university in Tubingen. His original aim was to become a Lutheran minister, but he also studied mathematics and astronomy, with a teacher who was one of the leading supporters of the Copernican view. Encouraged by his university teachers, Kepler's career quickly moved towards astronomy.

Kepler seems to have been convinced of the heliocentric model from an early age. He thought that it made it possible to *explain* the movements of stars and planets, rather than just predict them. He wrote, for instance, of the 'whirling effect' of the Sun keeping the planets moving in their orbits round it.

On 1 January 1600, Kepler fled from Austria to escape from conflict between Catholic and Protestant factions, and took a job as Tycho Brahe's assistant in Prague, where he was now based. When Tycho died rather suddenly in 1604, Kepler inherited his data. Shortly before his death, Tycho had set Kepler the task of working out the orbit of Mars. He boasted that he would do it in 8 days – in fact it took him 8 years. His problem was that he could not get Tycho's data to agree with the predictions of the Copernican model. They were out by 6 minutes of arc. This is a tiny amount, which many scientists would have ignored, but Kepler had total trust in the accuracy of Tycho's measurements and *knew* that the model wasn't right. After several years of painstaking calculations, trying to fit different kinds of circular orbits to the data,

Key terms

The usual unit for measuring angles is the **degree**. A full circle is 360°. To measure smaller angles, we use minutes and seconds.

1 minute is $1/60$ of a degree.

1 second is $1/60$ of a minute (or 1/3600 of a degree).

Sometimes we use the term 'a minute of arc' or 'a second of arc', to make clear that we are talking about an angle and not a time measurement.

Key terms

Imagine 'lining up' a lamp-post so that it looks exactly in line with a distant object. If you now move a step to the side, they no longer appear in line. The apparent motion of the lamp-post (relative to the distant object) is due to **parallax** – you have moved to a slightly different position, so the line from you to the lamp-post is in a slightly different direction.

The same principle can be used to measure the distance to the Moon – by measuring the difference in its direction when viewed from two places on the Earth's surface at the same moment. (Can you see why it has to be at the same moment?)

Questions

5 Tycho obtained evidence that contradicted aspects of the Ptolemaic model. What were the main pieces of evidence? How did they conflict with the Ptolemaic model? This evidence did not immediately lead all astronomers to reject the model. Why not?

6 What seems to have led Kepler to accept the heliocentric model?

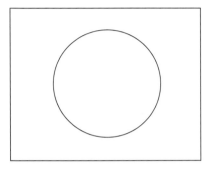

Figure 15.14

An ellipse looks like a slightly flattened circle. You should not think, however, that the orbits of the planets are pronounced ellipses. This is an accurate drawing of the shape of the orbit of Mars. It is an ellipse, but it looks like a circle. It is not hard to see why it took so long for Kepler to realise that the orbit was elliptical.

Discussion point

It was Kepler's work that really established the heliocentric model. Did Kepler collect any new data that supported this model? How did his work lead to the acceptance of the model?

he came to the conclusion that the orbits of the planets were not circles, but ellipses. This was an enormously daring leap, as circles had been the basis of models of the solar system for over 2000 years. Using elliptical orbits, Kepler drew up a large set of tables of predicted planetary positions (the *Rudolphine Tables*). Unlike the predictions of Ptolemy and Copernicus, these really did agree very closely with observation.

The statement that 'the orbits of the planets are ellipses' is known as Kepler's first law of planetary motion (Figure 15.14). Kepler also proposed two further laws:

- each planet moves round its orbit so that the line joining it to the Sun sweeps out equal areas in equal times;

- the time it takes for each planet to orbit the Sun (T) is related to its average distance from the Sun (R) by a simple rule: the number R^3/T^2 is the same for every planet.

Galileo Galilei: challenging the authority of tradition

Galileo Galilei was born in Pisa in 1564 (the same year as Shakespeare). In 1592 he was appointed Professor of Mathematics at Padua, one of the leading universities of the day. Galileo's early work on motion challenged the established view of Aristotle. As the majority of natural philosophers of Galileo's day were convinced Aristotelians, he created many professional opponents through this work.

Experiment not authority: the basis of modern science

Ancient Greek philosophy was based on reasoning. If you wanted to understand something, you read books by learned people, and thought hard about it. Galileo's view was that you had to carry out experiments to find out what really happens, and base your explanations on that. This seems an obvious approach to us today, but in Galileo's time it was revolutionary. Because he was the first to argue this way, Galileo is regarded as the 'father' of modern science.

Like Kepler, Galileo seems to have been convinced, from his early days, of the heliocentric view. He was also a sincere Catholic and became increasingly concerned that the Church was in danger of making a major error in supporting the geocentric view. At the time there was a rather uneasy truce between astronomy and the Church. Astronomers were allowed to use the Copernican model as a calculating device, but should not claim that it was a picture of how things really were. They were also to keep out of theological matters. Galileo was convinced that evidence would eventually show that the heliocentric view was correct, and the Church's support for the wrong view would fatally undermine its authority. He expressed his views very confidently, sometimes appearing rather arrogant. He was also a person who did not suffer fools gladly. He called some of his Aristotelian opponents 'dumb idiots'. As a result he antagonised opponents and made enemies of some of them. It particularly angered him that some of his opponents brought theological issues into the argument, rather than keeping it a purely philosophical matter.

In 1609 Galileo obtained one of the first telescopes. (They were invented in Holland around 1608.) He immediately saw how useful it could be and made a better one (which magnified 30 times) to observe the night sky. Between 1609 and 1615 he observed mountains on the Moon and sunspots (both showing that heavenly bodies were not 'perfect' spheres), moons of Jupiter (showing that something could orbit around a centre other than the Earth), and phases of Venus (Figure 15.15).

In 1615, following an after-dinner discussion with his patron, the Grand Duchess Christina of Tuscany, Galileo wrote her a long letter setting out his views on the solar system. In it he crossed an important line by arguing that the heliocentric model should be regarded as 'fact'.

He also discussed theological issues, arguing that the heliocentric view did not conflict with scripture. He argued that science and theology have different aims: 'the intention of the Holy Ghost is to teach us how to go to heaven, not how heaven goes'.

Galileo's opponents were quick to draw this letter to the attention of the authorities. Cardinal Roberto Bellarmine, leader of the Jesuits and probably the most powerful man in the Catholic Church at the time, censured Galileo, instructing him that the heliocentric model must be regarded as a hypothesis until there was evidence for the Earth's movement. Copernicus's model should be thought of as a calculating tool, which 'saves the appearances'; it does not claim to tell us how things are. Galileo went to Rome in 1616 to clear himself of these rumours of heresy and to argue his position. He was cleared, but Bellarmine's instructions were also confirmed. In fact Galileo seems not to have been too put out by this, believing that he could collect convincing evidence that would show that the Earth moved.

Question

7 How did Galileo's observations provide evidence against the geocentric model? Did any of them provide evidence for the heliocentric model?

Figure 15.15

The phases of Venus (similar to the phases of the Moon) showed that Venus was sometimes on the opposite side of the Sun (as the heliocentric model predicts). If it was always between the Earth and the Sun (as the geocentric model predicts), we would never see it 'full'.

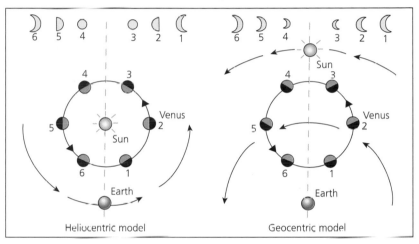

Technology and science

Sometimes advances in scientific knowledge lead to new technical developments and new devices. But it can also work the other way round. The telescope was developed by craftsmen, before the scientific principles it is based upon were understood. It made it possible to see things in the sky that are invisible to the naked eye, and so enabled our understanding of the solar system to advance.

By 1630, Galileo was ready to write his great work, the *Dialogue Concerning the Two Chief World Systems*. He believed his clinching evidence of the Earth's movement came from his theory of the tides. (Strangely he did not make much use of Kepler's work, though he must

8 Explain what it means to treat the heliocentric model as a hypothesis which 'saves the appearances'.

9 Some powerful figures in the church opposed Galileo's efforts to spread the heliocentric view. Why do you think they were opposed to it? What was Galileo's response?

10 The title of Galileo's book refers to 'the Two Chief World Systems'. Explain what these were.

11 What advantages were there for Galileo in writing his book in the form of a dialogue?

Discussion point

'The Galileo affair tells us more about political power and rivalry, than about the relationship between science and religion.' List the evidence which supports this statement, and that which casts doubt on it. What is your own view?

In what way was Galileo's book a 'milestone in the struggle for freedom of thought and expression'?

have known of it. And we now know that his theory of the tides was completely wrong.) Rather than write a formal treatise in Latin, he chose to write in Italian (which everyone could read) and to construct the book as a dialogue between three characters. Galileo puts his own views into the mouth of Sagredo, while Simplicio is a caricature of the Aristotelian philosophers who opposed Galileo. Salviati is open-minded and willing to listen to all arguments. The reaction of the Church was rapid; the printer was ordered to cease printing and all copies were seized. Galileo, now almost 70, was summoned to stand trial in Rome before the Inquisition. Although now ill and infirm, Galileo was examined by the inquisitors in April 1633. Only after being shown the instruments of torture (which could not by law have been used on a man over 70), did he give way. Dressed in the white robes of a penitent, he was required to kneel and publicly renounce his Copernican views. Galileo remained under house arrest for the last years of his life, cared for by his daughter, but forbidden to teach or publish.

The episode ended a highly creative period in Italian science. Four centuries later, Pope John Paul II acknowledged that the Church had been wrong to deal with Galileo as it did. But what exactly went wrong? What was the dispute really about? It is often portrayed as a clash between science and religion. But Galileo was an ardent Catholic all his life. In part it was about authority: if there is a conflict of ideas, should the final decision be based on evidence or on tradition? The Church was clearly threatened by the challenge to its authority, and wanted to draw clear boundaries between what could and could not be questioned. Over time, however, this has proved to be an impossible line for both church and secular authorities to draw. Another key aspect of the Galileo case lies in the personalities involved. In 1616, Cardinal Maffeo Barberini was one of Galileo's supporters. By 1632, he had become Pope Urban VIII. Galileo made the serious tactical error of putting one of Urban's favourite arguments into the mouth of Simplicio, thus making it look naïve and foolish. This may have been one of the reasons for the Church's reaction to the *Dialogue*, though internal feuds in the Vatican seem also to have played a part. Historians still argue about motivations of the main characters involved in the episode.

With the benefit of hindsight we can now see that the *Dialogue Concerning the Two Chief World Systems* is one of the most important books ever written. It is a classic of science, and of literature, and a milestone in the struggle for freedom of thought and expression.

An explanation for the planetary motions

Kepler had proposed a model of the solar system that fitted the data. But could the motions described by the model be explained? Galileo had also done some important work on motion. Isaac Newton, who was born in 1642, the year that Galileo died, built on it to develop his famous Laws of Motion.

In 1664, just after passing his examinations at Cambridge, Newton came home to Lincolnshire to escape from the plague. This gave him time to think about questions that interested him. One was: why does the Moon go round in an orbit, unlike objects near the Earth which simply fall? He argued that the Moon *is* falling towards the Earth – and this keeps it in its orbit, rather than moving off in a straight line. The force making the Moon 'fall' is the same gravitational attraction that makes objects fall on Earth. Gravity does not act only on objects on Earth – it is a universal force between any two objects.

To check this idea, Newton worked out how strong the gravitational force would be at the distance of the Moon, as compared with its strength at the Earth's surface. The Moon is 60 times as far from the centre of the Earth as an object near the Earth's surface (384 000 km compared with 6400 km). If the gravitational force falls off as the square of the distance (an inverse-square law), it will be 1/3600 times weaker at the Moon than it is on Earth. So the Moon should 'fall' 1/3600th as far every second as an object on Earth. When Newton did the calculation, he had an inaccurate value of the radius of the Moon's orbit – and the figures did not tally exactly. So he did not immediately publish his work. In fact it was not published until 1687, in Newton's most famous book, the *Principia*.

One reason why Newton proposed an inverse-square law of gravitation was that it led straight to Kepler's three laws – they could be deduced from it. It explained why the planets move as they do.

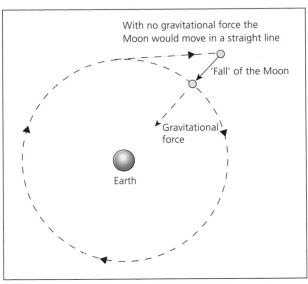

With no gravitational force the Moon would move in a straight line

'Fall' of the Moon

Gravitational force

Earth

Figure 15.16

The fall of the Moon.

The idea of a field

Newton's idea of action at a distance was very controversial. It seemed rather like magic – two objects exerting an influence on each other across empty space. There are other examples of forces like this which act at a distance. One is magnetism; another is the force between electrically charged bodies. Scientists have proposed the idea of **fields** to help explain these forces. There is a **gravitational field** around every mass. If another mass comes into the field, it experiences a force. The more massive an object, the stronger the field around it. The field also gets weaker as you go further from the object.

Discussion point

Newton's work did not lead to better predictions of the positions of the planets. In what way, therefore, could it be said to have improved our understanding of the solar system?

Coping with a challenge

As far as all astronomers before 13 March 1781 were concerned, the solar system comprised the Sun and six planets. Then, quite unexpectedly, an amateur astronomer in Bath, working with a home-made telescope, discovered Uranus. William Herschel was looking for comets at the time. Some thought his discovery a matter of luck, but in fact he was engaged in a systematic and detailed survey of the sky. Herschel had a fantastic visual memory (remember that there was no photography in 1781) and this enabled him to spot changes in the

12 What method would Herschel
have used to estimate the
distance to Uranus? The
measurements took him several
weeks. Why was this?

13 Why do you think Adams (and
others) were reluctant to reject
the accepted model of Kepler
and Newton? What did they do
instead?

14 Explain why the anomalies
observed in the orbit of Uranus
posed such a serious threat to
the accepted understanding of
the solar system.

patterns of stars. He estimated the distance to Uranus and found it was far beyond Saturn. At a stroke he had doubled the size of the solar system.

As other astronomers observed Uranus, however, they began to encounter difficulties in working out the details of its orbit and predicting exactly where it should be at different times. It seemed to stray off its predicted course. This was a serious challenge to the accepted model of Kepler and Newton, which was believed to be highly accurate. Some people, including the Astronomer Royal, George Airy, thought that Newton's Law of Gravitation might not apply at longer distances. Others suggested explanations that were compatible with the accepted model: perhaps Uranus had an invisible moon, or was struck by a comet just before it was discovered.

In 1843 John Couch Adams had just completed his degree at Cambridge in mathematics – with double the mark of the next best student in his year. Adams thought that the irregularities in the motion of Uranus might be caused by another planet, even further away (Figure 15.17). After 2 years of painstaking calculations, he wrote to the director of the Cambridge Observatory, John Challis, in 1845 predicting where this new planet should be. Had anyone looked they would have found it, within 2 degrees of the position that Adams predicted. But no one did!

Instead Challis sent him off with a letter of introduction to the Astronomer Royal at Greenwich. Airy was in France so Adams returned a month later again without making an appointment – this time Airy's butler told him that Airy was at dinner and could not be disturbed. Adams left a copy of his calculations and went back to Cambridge. When Airy eventually got round to looking at these, he sent Adams some technical questions, relating to his own idea that Newton's Law of Gravitation broke down at large distances. Adams' earlier treatment had made him rather fed up with Airy and so he didn't reply – and Airy then assumed his questions had shown up flaws in Adams' calculation.

Meanwhile a French astronomer, Urbain Leverrier, was working independently on a similar calculation. By June 1846 he too could predict the position of the unknown planet. But like Adams he could not persuade any French astronomers to look. However, his paper was noticed by Airy who compared the prediction with Adams's and found they were almost identical. This made him think it was worth looking. Ironically the best telescope was Challis's, in Cambridge, so he asked him to start a search. Instead of going straight to the predicted point, Challis embarked on a rather slow and painstaking search of a large area around it. This would certainly find the planet if it existed, but was slow. Twice in August Challis actually observed the unknown planet – but did not recognise it. On 29

Figure 15.17

The gravitational attraction of another (as yet undiscovered) planet beyond Uranus could explain the 'wobbles' in its motion.

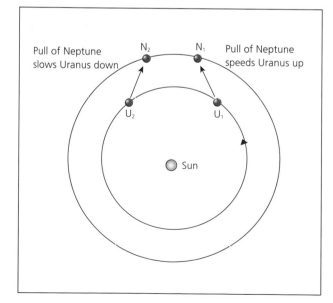

Pull of Neptune slows Uranus down

N_2 N_1

Pull of Neptune speeds Uranus up

U_2 U_1

Sun

September he found one 'star' that he thought might be the planet, but decided to wait till the next night before looking at it again with higher magnification. By the next night, the skies were clouded over.

In any case, he was already too late. On 23 September, a letter from Leverrier arrived at the Berlin Observatory. The assistant astronomer, Johann Galle, persuaded the director to begin a search immediately. With a telescope, a planet shows up as a disc, while a star remains just a point of light. Galle hoped to see that one of the 'stars' in the predicted region was a disc, but was initially disappointed. Then a student present suggested they check off the stars they could see against a recently published chart of that region of the sky. Checking through them one by one, they found a new 'star' that was not on the chart. In one night they had done what Challis had failed to do in 2 months; they had found the unknown planet.

Leverrier was generally acclaimed as the discoverer of a new planet, now named Neptune. Although he knew Adams had predicted it first, Airy did nothing to get Adams the credit he deserved. However, John Herschel (the son of William and also a leading astronomer) made Adams' contribution known and arranged for Adams and Leverrier to meet at his house in Kent. Both are now recognised as the joint discoverers of Neptune.

As a result of this episode, people were even more convinced that the heliocentric model of Kepler, together with Newton's idea of universal gravitation was right. It had successfully withstood a serious challenge.

An understanding that works!

Perhaps the most convincing way to show that your understanding of anything is correct is to make something complicated that depends on it, and show that it works. The *Voyager* missions certainly showed that our understanding of the solar system works. They are among the most remarkable technical achievements of all time.

The *Voyager* missions were planned by scientists at NASA (the US National Aeronautics and Space Administration) to take advantage of a rare arrangement of the four outer planets in the late 1970s and early 1980s. All four happened to be on the same side of the sun, making it possible for a spacecraft to follow a path which would pass close by each. Even better, the gravity field of each planet could be used as a 'slingshot' to accelerate the spacecraft on its way. *Voyager 1* and *2* were both launched in the summer of 1977. *Voyager 1* passed close by Jupiter and Saturn. *Voyager 2* made an even more amazing journey, passing close to all four outer planets, and several of their moons (Figure 15.18). It took 12 years and travelled 7128 million kilometres to reach Neptune. On 25 August 1989, it passed just 5000 km from the surface of Neptune, within 100 km of the planned distance. This is equivalent in accuracy to sinking a golf putt of 3630 km!

Discussion points

Anomalies in the motion of Uranus did not make astronomers immediately give up the accepted heliocentric model, and Newton's Law of Gravitation. Were they being good or bad scientists in holding on and hoping that they would eventually be able to explain the anomaly?

What seems to be necessary to make scientists give up a well-established idea? (You might be able to find a few examples in other chapters of this book.)

Questions

15 Galle found Neptune by spotting a 'star' that was not on his chart. What other method might he (or Challis) have used to identify a planet?

16 Whereas the anomalies in the orbit of Uranus had made astronomers question the accepted model of the solar system, the discovery of Neptune now made them even more confident in their model. Explain why it had this effect.

Discussion point

Do you agree that the most convincing way to show that your understanding of anything is correct is to make something complicated that depends on it and show that it works? Can you think of any other examples – either from elsewhere in this book, or from your general knowledge?

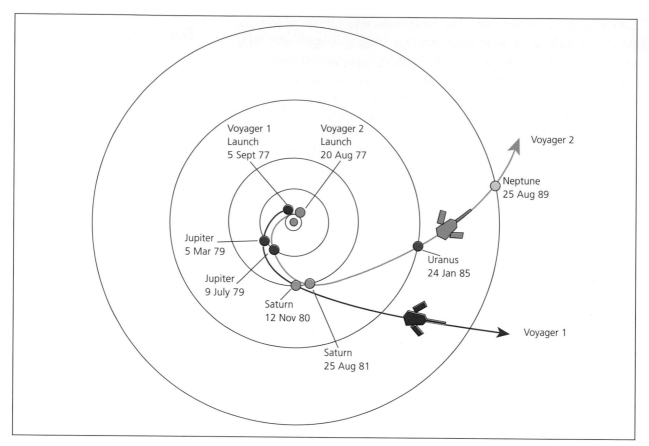

Figure 15.18

The journeys of Voyager 1 and 2.

Discussion point

The total cost of the Voyager missions was US$865 million. Is this a good way for a wealthy country like the USA to spend its money? What points would you make for and against?

The *Voyagers* sent back some remarkable and beautiful photographs of the outer planets and their moons. Biologist and author Stephen J. Gould wrote the following about *Voyager 2* and one of the photographs it took (Figure 15.19):

Knowledge and wonder are the dyad of our worthy lives as intellectual beings. *Voyager* did wonders for our knowledge, but performed just as mightily in the service of wonder – and the two elements are complementary, not independent or opposed. The thought fills me with awe – a mechanical contraption that could fit in the back of a pickup truck, travelling through space for twelve years, dodging round four giant bodies and their associated moons, and finally sending exquisite photos across more than four light-hours of space from the farthest planet in our solar system. [At the time, Neptune was further away than Pluto.] As *Voyager* passed Neptune, her programmers made a courtly and proper bow to aesthetics and took the most gorgeous picture of all, for beauty's sake – a photograph of Neptune as a large crescent, with Triton as a smaller crescent at its side. ... Let it stand as a symbol for the fusion of knowledge and wonder.

Stephen J. Gould, Bully for Brontosaurus, *Penguin, 1991*

17 To mark the millennium, a group of people in York made a scale model of the solar system alongside a long straight cycle track. (See: www.solar.york.ac.uk). If a group in your school, or college, or neighbourhood, wanted to do something similar, what advice would you give them on the sizes to make the Sun and planets, and the distances between them? Write your recommendations in a report.

Ref to page 193

18 Construct a time-line diagram to show the important steps in our growing understanding of the solar system.

19 If you wanted to persuade someone today that the Earth goes round the Sun, rather than the other way round, what evidence would you use to try to convince them?

20 Which was more important in the development of our understanding of the solar system: collecting more accurate data, or creative conjectures about how to explain the data?

Figure 15.19

Neptune and its moon Triton as seen from Voyager 2. A stunning photograph made possible by equally stunning science and technology.

16

Understanding the universe

The issues

As our understanding of the universe has grown, the Earth has been steadily pushed away from the centre of things. We have also had to come to terms with the evidence that the universe is vast, with almost unimaginable distances between the stars and the galaxies, and with the idea of an expanding universe, that began in a 'big bang' over 15 billion years ago. These are disorienting ideas, if we take them seriously. They certainly put our concerns about matters on Earth in context! Yet the desire to satisfy our curiosity about the universe is a very fundamental expression of our humanity – and we continue to spend large sums of money on astronomy even though the findings can have little direct impact on the quality of material life on Earth.

The science behind the issues

The nearest star to our Sun is over 4 light years away – around 1000 times the radius of the solar system itself. Most stars are much further away. The Sun is towards the outer edge of a spiral galaxy, the Milky Way. This, in turn, is just one of millions of galaxies which make up the universe. The distant galaxies are moving away from us at high speeds. The universe began in a 'big bang' around 15 billion years ago.

What this tells us about science and society

The growth of our understanding of the structure and size of the universe depends on accurate measurements and observations. However, astronomers are often working at the limits of their instruments and so all measurements are estimates which may have substantial errors. Some advances were only possible when better telescopes and new techniques like spectroscopy were available.

Patterns and explanations did not simply 'cmcrgc' from the data but were the result of imagination and conjecture. As a result, there have been debates about the data and how to interpret it, and estimates of distance have often had to be revised. The 'big bang' theory became widely accepted because it was predicted by the General Theory of Relativity and supported by evidence that the galaxies are moving away from us. Confidence in the theory grew when its microwave 'echo', predicted by theory, was detected in 1965.

Figure 16.1

The Hubble space telescope: a powerful tool for exploring the universe.

How far are the stars?

During the 1640s and 1650s, the heliocentric model of the solar system gradually took hold as the accepted view. The simplicity of Kepler's model, based on elliptical orbits – and the very accurate predictions it made – were what convinced astronomers. Ref to page 200

If the Earth is moving in a large circle round the Sun, then we would expect to see changes in the patterns of stars at different times in the year (Figure 16.2). When Copernicus first proposed his heliocentric model, astronomers looked for this *stellar parallax* but found none. Some saw this as evidence against the heliocentric model – others that it showed how far away the stars really were! Now, with the heliocentric model established, most people accepted the second explanation and the search for stellar parallax was on.

Key terms

Stellar means 'of the stars'. So **stellar parallax** means parallax of the stars.

Figure 16.2

A star which is closer to us should move slightly, against the background of very distant stars, because we are looking at it from a different position. The effect is called parallax. If we can measure the change in angle of the closer star, then we can estimate its distance, as we know the distance between the two viewing positions (it is equal to the diameter of the earth's orbit).

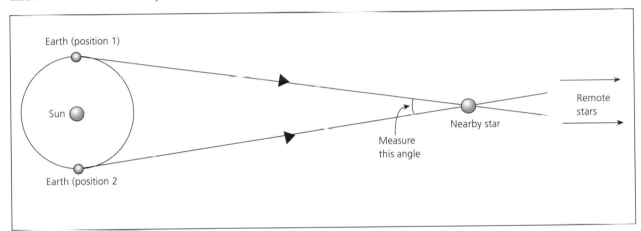

Parallax observed!

One of the problems in searching for stellar parallax is to know which stars to look at. Which ones are closest to us? One guess is that the closest ones will be bright. Another clue is the fact that some stars actually do move slightly relative to their neighbours – something called 'proper motion'. Maybe they all move but we only notice it for the closest ones. Several astronomers used these 'hunches' to pick out good stars to observe closely. In 1838, Friedrich Wilhelm Bessel succeeded in measuring a parallax angle of 0.31 seconds for the star 61 Cygni (Figure 16.3). From this, he calculated that its distance is 11.2 light years.

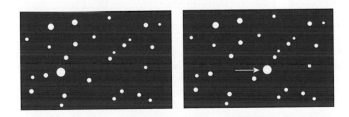

Figure 16.3

These two images, from a modern telescope, show stellar parallax. The bright star in the lower left-hand part of the first photo has moved to the right. By measuring the angle it has moved through, relative to the more distant stars in the background, we can work out how far away it is.

Shortly afterwards, several other astronomers measured similarly small parallax angles for some other stars. The closest stars are over 4 light years away – about 750 times larger than the diameter of the solar system. Our solar system is like a tiny island in empty space.

Measuring tiny angles

Figure 16.4

To get a feeling for how small these angles that astronomers were trying to measure actually are consider the dot below. At a distance of 1 kilometre this would have an angular width of one second.

Bessel measured an angle which is about 1/3 of this. This was only possible because of the improvements in telescopes by the 1830s. With modern telescopes we can measure parallax angles down to about 0.01 second.

New techniques

Two developments in the 19th century played a key role in the growth of knowledge of the universe. The first was the invention of *photography*. The first photographs were taken around 1827, though the process only became widely available in the 1840s and 50s. As photographic techniques improved, astronomers began to see its potential. The great advantage was that observations could be recorded and studied later. This made it easier to spot changes. Also, by using longer exposure times, it was possible to study objects that were too faint to be observed in 'real time', even with a telescope.

The second development was *spectroscopy*.

Spectroscopy

Figure 16.5	Figure 16.6

The spectrum of sunlight, showing the dark lines observed by Fraunhofer.

The emission spectrum of hydrogen produced when the atoms in the gas are excited by heating or by electricity. The lines are a unique fingerprint of the element.

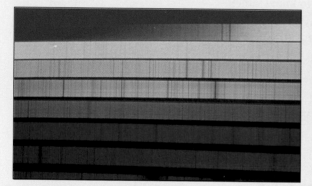

The scientific study of spectra began with Isaac Newton. Using a prism, he demonstrated that sunlight is made up of a continuous band of colours, from red to violet. Around 1800, Thomas Young obtained evidence that persuaded scientists that light is a wave, and that light of different colours has different wavelengths, from red (shortest) to violet (longest).

In 1814, Joseph Fraunhofer noticed that the spectrum of the Sun is crossed by many dark lines (Figure 16.5). Meanwhile, two other German chemists, Robert Bunsen and Gustav Kirchhoff, had found in the laboratory that the spectrum of a glowing vapour consists of bright lines of different colours. The pattern of lines is characteristic of the element present in the vapour (Figure 16.6). The spectrum of a glowing hot solid object, on the other hand, is a continuous band of colour. Kirchhoff realised that the lines Fraunhofer had seen in the Sun's spectrum were due to light from the Sun's surface passing through the slightly cooler vapour surrounding it – and that he could use this to discover the composition of the Sun. In 1859 he published his findings: the Sun is composed mainly of hydrogen and helium, and the vapour surrounding it contains other heavier elements found on Earth. So we had evidence for the first time that celestial objects are made of the same material as the Earth.

Beyond the closer stars

Measurements of stellar parallax depend on the fact that some stars are much closer than others. In fact only a few stars were close enough for their parallax to be measured. Most were so distant that no parallax could be detected. As telescopes have improved, this number has increased, but we can still only detect parallax for a tiny fraction of all the stars we see. How then can we measure the distance to the others, and to the other objects we see in the night sky?

By the late 19th century, no one could see how to do this. There is not all that much, after all, that you can observe about a star. You can measure carefully its position in the sky, you can measure how bright it is, and (with spectroscopy) you can measure the composition of its light. But you cannot work out from that how far away it is. One star might be dimmer than another because it really is less bright, or because it is further away.

One group, at the Harvard College Observatory, decided to compile a catalogue of stars, with their spectra. This was a huge undertaking, involving hours of painstaking work, with no obvious goal other than to collect reliable data. The work was deemed to be too routine for a professional astronomer, so the director of the Observatory employed a team of women, who were not permitted at that time to obtain a degree at Harvard, let alone be on the staff. Between 1890 and 1924 they steadily built up a huge catalogue of spectra of stars, eventually listing 225 000 stars. One member of the team, Annie Jump Cannon, developed earlier ideas about classifying stars according to their spectra. She began with seven classes, ranging from bluish stars, through yellow and orange, to red. With subsequent improvements and additions, this rose to 13. In 1914, using the data in the catalogue, an American astronomer, Henry Norris Russell, tried plotting the brightness of a star against its spectral class. If you do this using the apparent brightness, there is no pattern. But if you take account of how far away the star is, and work out its actual (or intrinsic) brightness, then a pattern does emerge (Figure 16.7). Of course, it is only possible to do this for stars whose distances are known, from parallax measurements.

This was a significant breakthrough. For it means that, by observing the spectrum colour of a star, astronomers can estimate its intrinsic brightness. By comparing this with its apparent brightness, they can then work out how far away it is. This allows them to estimate distances to stars that are too far away to be measured by parallax.

Cepheid variables

The work of one woman in the Harvard team led to a discovery that was to prove particularly important. Henrietta Swan Leavitt was studying stars in one of the Magellanic Clouds. These are two clusters of stars that are visible in the Southern Hemisphere. Leavitt's data was in the form of photographic plates from the telescope at Harvard University's Observatory in Peru. Leavitt was interested in variable stars known as

Questions

3 Why can you not work out the distance to a star by measuring how bright it looks?

4 Explain why a star that is far away from us will look dimmer than a similar star that is closer.

Key terms

Intrinsic brightness is a measure of how bright a star really is. **Apparent brightness** means how bright it looks to us. The two are not the same, because stars are not all the same distance away. A faint star might have a low intrinsic brightness; but it could also be an intrinsically bright star which is very far away.

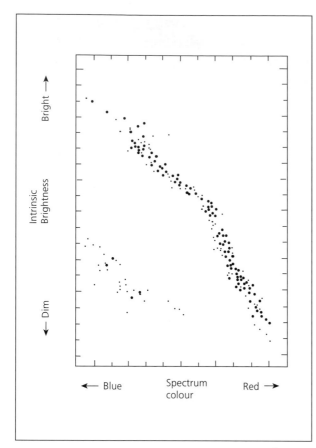

Figure 16.7

This graph is known as a Herzsprung-Russell diagram. Most stars lie on the 'main sequence' running from top left to bottom right.

Key terms

Most stars are constant in brightness. But some vary in brightness, for different reasons. These are called **variable stars**.

One group of variable stars rises and falls in brightness in a steady regular pattern. The first of these to be noticed was Delta Cephei (in 1784), and so the type are known as **Cepheid variables**.

The time it takes for a full cycle from maximum brightness to minimum, and back to maximum again, is called the **period** of the star.

Figure 16.8

Henrietta Leavitt at work.

Cepheid variables. These stars pulse in brightness, a bit like the light from a lighthouse. The brightness rises and falls over several days. Leavitt listed 1777 variable stars, and measured the periods of 16 of these. Then, in 1912, she noticed a pattern: the brighter a variable star, the longer its period. She was able to see the pattern because all the stars in the Magellanic Cloud are roughly the same distance away – so their apparent brightness corresponds to their intrinsic brightness.

Leavitt's discovery is important because it means that you can calculate the intrinsic brightness of any Cepheid variable if you know its period. Then by measuring its apparent brightness, you can deduce how far away it is. There is just one snag: you need to know the distance of *one* Cepheid variable to get started. But in fact no Cepheid variable is close enough for its distance to be measured by the parallax method. However, the astronomer Harlow Shapley, who was director of the Harvard Observatory from 1921, used a number of additional assumptions to estimate the distance to one Cepheid – which then allowed the distances to all other Cepheids to be estimated.

Question

5 Explain why Leavitt would not have obtained any pattern linking period with brightness if she had worked on Cepheid variable stars in general. Why was it important that she used stars in the Magellanic Cloud?

The structure of the universe

As well as the stars we observe in the night sky, there are also many fuzzy, or cloudy objects, which astronomers call *nebulae*. Some are visible with the naked eye. The most prominent is a hazy band that runs across the sky, called the Milky Way. The Magellanic Clouds, visible only from the southern hemisphere, are two more. Another smaller one is visible in the constellation of Andromeda. These had puzzled astronomers for centuries. One view was that they were clouds of gas and dust, relatively close to us. Another was that they were clusters of stars, a very long way away.

An early supporter of the second view was the philosopher Immanuel Kant. He speculated (correctly as it turned out) that the Milky Way was really our edge-on view of our own galaxy – and that the nebulae are other galaxies beyond our own. In 1755 he wrote that 'the nebulae are systems of stars lying at immense distances'. He called them 'island universes'. This was, however, just armchair theorising, and the question of what nebulae are was to rumble on for another 160 years.

William Herschel, who discovered Uranus in 1781, also observed and documented nebulae of different kinds (Figure 16.9). Some appeared to be stars surrounded by a halo of gas, others he took to be distant clusters of stars. He proposed a classification of five types. Later discoveries led to changes in this, but Herschel's basic idea was right – the nebulae are not all the same kind of thing. The development of spectroscopy reinforced this conclusion: some nebulae have spectra like that of a star, whilst others have spectra characteristic of a glowing gas. At the beginning of the 20th century, the common view among astronomers was that nebulae are stars at different stages in the process of formation. Ref to page 219

Question

6 Why do you think astronomers throughout the 19th century were not sure about what nebulae were? What evidence might have resolved the issue?

Figure 16.9

(a) An optical image of the Orion nebula which is at a distance of about 1500 light years.
(b) A processed image of the ring nebula in the constellation of Lyra recorded by a telescope with a light-sensitive silicon chip. The computer processing of the data gives a colour coded image according to the brightness of the light. White represents the highest intensity. The nebula is a shell of gas ejected from a central star (the yellow and blue dot at the centre). The distance to this nebula is over 2000 light years.

(a)

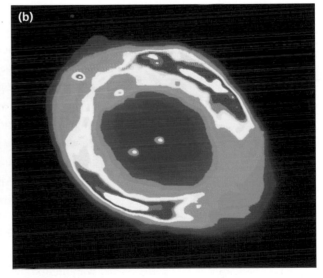

(b)

Leavitt's discovery of the link between the period of Cepheid variables and their intrinsic brightness provided a tool for tackling this problem. Telescopes allowed separate stars in some nebulae to be observed, and some of these were Cepheids. With great enthusiasm, the American astronomer Harlow Shapley embarked on a programme to measure distances to many of these star clusters. In just a few years, he had made some remarkable discoveries. He found that the clusters seemed to form a huge sphere, whose centre was far away in the direction of the constellation Sagittarius (visible from the southern hemisphere). In a daring imaginative leap, he conjectured (and astronomers still think he was right) that the centre of these clusters was the centre of the galaxy – and hence that the Sun was nowhere near the centre (Figure 16.10). Just as Copernicus and Kepler had made us accept that the Earth is not at the centre of the solar system, we now (in 1920) had to recognise that the Sun and the solar system are not at the centre of the galaxy.

Figure 16.10

Astronomers believe the Milky Way is a huge disc of stars, with a central bulge, surrounded by some globular clusters of stars. The Sun is towards the outer edge of the disc. The galaxy is thought to have a spiral structure, though the evidence for this is not yet conclusive.

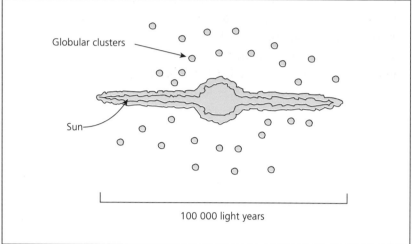

Globular clusters

Sun

100 000 light years

Question

7 Would it be correct to say that Shapley had *proved* that the centre of the galaxy lay in the direction of Sagittarius and that the Sun was nowhere near the centre? Explain your answer.

Not all of Shapley's conclusions turned out to be right, however. His distance measurements, using the Cepheid variable method, were all considerable over-estimates. (The reason for this was only discovered much later and need not concern us here.) As a result, his estimate of the size of the Milky Way galaxy was more than three times the size we now believe it is (around 100 000 light years). Because he thought the Milky Way was so large, he concluded that it was, in fact, the whole universe and so everything was within it – including all the nebulae.

Not everyone agreed. The issues came to a dramatic head at a debate at the National Academy of Sciences in Washington on 26 April 1920, between Shapley and another American astronomer, Heber Curtis. Curtis argued for the 'island universes' view – that some nebulae were galaxies, far beyond the Milky Way. The issue was resolved not by debate, but by evidence. The astronomer who collected this evidence was Edwin Hubble.

Island universes

In October 1923, Hubble was studying the Andromeda nebula (as it was then known) (Figure 16.11). The magnification of his telescope was sufficient to enable him to detect a few individual stars in the nebula. In one pair of photographs taken a few nights apart, Hubble noticed that one of these stars was varying in brightness – and a few more observations told him it was a Cepheid variable. That meant that he could use Leavitt's results to work out its intrinsic brightness, and from that he could work out how far away it was. His answer was an astonishing one: the star was 900 000 light years away – many times further than any star in our galaxy. Over the next year, he found some more Cepheid variables and checked his calculations. The results were

Figure 16.11

The Andromeda galaxy. Our present estimate is that it is 2.2 million light years away (more than double Hubble's estimate). Despite this huge distance, it is our nearest neighbour among the galaxies. It is one of the 'local group', along with our galaxy, the Milky Way. It is the most remote object you can see with the naked eye. If anyone asks you how far you can see on a clear day (or night), the answer is 2.2 million light years.

consistent. The Andromeda nebula was really the Andromeda galaxy – another 'island universe', far beyond our own galaxy. Hubble's results were reported to the American Association for the Advancement of Science on New Year's Day 1925. Everyone present knew the debate was over. Our galaxy is not the whole of the universe. It is one of many galaxies that make up the universe.

An expanding universe

At a meeting of the American Astronomical Society in 1914, well before the nature of nebulae had been sorted out, an American astronomer, Vesto Slipher, reported an unexpected discovery. He had found evidence that several spiral nebulae seemed to be travelling away from us at incredible speeds – up to 600 miles per second. The evidence came from measurements of their spectra, and a phenomenon called the Doppler effect.

The Doppler effect

Figure 16.12

The apparent wavelength of the waves is shorter when the sound source is approaching you, and longer when it is going away from you.

Almost everyone has noticed the way the sound made by a moving object changes when it passes you. A train whistle or horn shows the effect; so does the siren on a police car or an ambulance. As the vehicle approaches you, the sound is higher in pitch, and as it goes away from you, the pitch falls. The reason is that sound is a wave motion. As the vehicle approaches you, this makes the wavelength of the sound it emits 'bunch up'. A shorter wavelength means a higher frequency and so a higher pitch. Once it has passed and is going away, the sound waves are 'stretched out' a little. Longer wavelength means lower frequency so a lower pitched sound. The effect is named after the Austrian physicist who first explained it (in 1842), Johann Christian Doppler (Figure 16.12).

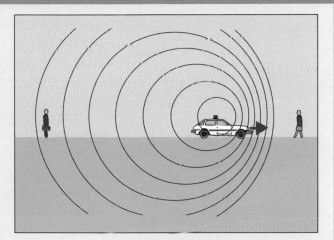

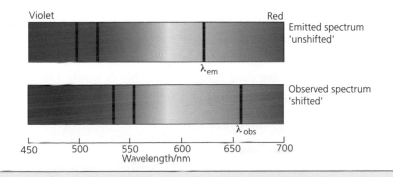

The Doppler effect also applies to light waves. If the source is moving towards you, the wavelength of the light is decreased. This makes it appear more blue than usual. If it is moving away, the wavelength is increased, making it look redder. If the light is from a glowing gas, its spectrum will be a series of lines. Coming from a receding object, these will have longer wavelengths than the lines in the spectrum of the same element in the laboratory. In other words, they show a *redshift* (Figure 16.13).

Figure 16.13

If a luminous object is moving away, its spectrum will be shifted towards the red end.

The reason for the redshifts Slipher observed were a mystery at the time. Others, however, confirmed his observations. And Slipher himself measured recessional velocities for more galaxies, including one of 1100 miles per second.

Around the same time, astronomers were arguing over the implications of Einstein's General Theory of Relativity. Published in 1915, this suggested that gravitation is due to the curvature of space by all objects that have mass. One prediction of the theory is that light should bend as it passes a massive body. This was observed by looking at stars close to the Sun during a total eclipse in 1919. The amount of bending was exactly as Einstein had predicted.

Einstein was aware that his theory had implications for cosmology. His equations led to the conclusion that the universe must either be expanding or contracting. He was unhappy about this and modified the equations to produce the result that the universe is static – as Einstein believed it must be. He was later to refer to this as 'the biggest blunder of my life'. In 1922, a young Russian meteorologist, Aleksander Friedmann, published a paper which showed that Einstein's original equations led naturally to the conclusion that the universe is expanding at a changing rate.

The evidence for this expansion was provided by Edwin Hubble. Using the same method that he had used to measure the distance to the Andromeda galaxy (based on Cepheid variables), Hubble estimated the distance to some of the other closer galaxies. He was not able to use the method for galaxies further away because the Cepheid variables in them were too dim to see clearly. He could, however, see the brightest stars in these more remote galaxies. So he used a roundabout method. First he worked out the intrinsic brightness of the brightest stars in each galaxy. This enabled him to confirm that the brightest stars in every galaxy had roughly the same intrinsic brightness. He then observed the brightest stars in the more remote galaxies and assumed that they too would have the same intrinsic brightness as those in the nearer galaxies. By comparing this with their apparent brightness he was able to work out how far away they were.

Armed with this information, Hubble then measured the red shifts of these galaxies to calculate the speed with which they are moving away from us. When he plotted this speed against the distance, he found that there was a pattern (Figure 16.14). The further away the galaxy was, the faster it was moving away from us. This, of course, means that the distance to other galaxies can be estimated by measuring their red shift, and then reading their distance off Hubble's graph. The bigger the red shift, the faster it is receding and the further away it is.

A point to note

Although the more distant galaxies are receding from us, the closer ones in the local group are actually approaching us, due to gravitational attraction. The Andromeda galaxy has a blue shift, rather than a red shift.

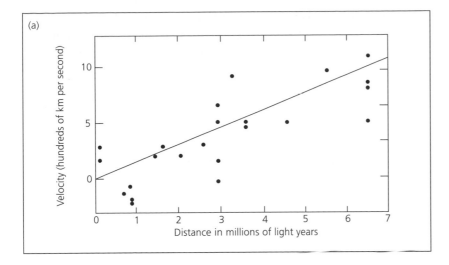

(a)

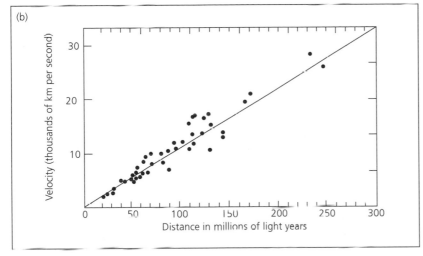

(b)

Figure 16.14

(a) *Hubble's measurements of distance and velocity for the galaxies he could measure;* **(b)** *a distance and velocity graph for more distant galaxies, based on modern measurements.*

Questions

9 Why do you think the points in Hubble's graph (Figure 16.14 (a)) do not all lie exactly on the line? (Hint: Consider the assumptions Hubble had to make to estimate the distances and speeds.)

10 Was Hubble being a 'good scientist' when he claimed there was a relationship between distance and speed?

11 By measuring the redshift of a galaxy, we calculate that it is moving away from us at a speed of 15 000 miles per second. Use Figure 16.14(b) to work out the distance to this galaxy.

12 Your uncle has read that the universe is expanding and asks you how we know this. How would you explain it to him?

A big bang

Hubble's work (and that of other astronomers like Slipher) showed that the universe is expanding. This suggests that the universe began with a big bang. At first, some astronomers did not like this idea because it implies a 'creation' at one moment in the past. It also implies that the first stage in the evolution of the universe cannot be explained by the normal physical laws. So some astronomers argued instead for a 'steady state' theory, with new matter being produced all the time as the universe expands. In this way, the appearance of the universe would remain constant, even though it is expanding.

One problem with the big bang theory was its prediction of the age of the universe. Using Hubble's measurements of the speed at which galaxies were moving apart, and his measurements of their distances, he was able to work backwards and estimate how long ago the big bang occurred. Hubble's answer was 2 billion years. However, in 1929, the estimate of the age of the Earth was 3.4 billion years. It is not possible for the Earth to be older than the universe. This anomaly was not resolved until 1952, when it was shown that Hubble had, in fact, observed a different type of Cepheid variable from those studied by

Henrietta Leavitt. This led to a correction in his estimates of distances to the galaxies – which doubled the estimate of the age of the universe to 4 billion years. Later work has increased this estimate further, to the present value of around 15 billion years.

The 'echo' of the big bang

Astronomers who championed the big bang theory during the 1930s and 40s tried to deduce some of its theoretical implications. They showed that Einstein's Theory of General Relativity implied that the universe had initially been confined to a small space, perhaps even an infinitely small pinpoint, or 'singularity'. Another implication was that there should be an 'echo' of the big bang still around – in the form of radiation from the bang itself. Because of the expansion of the universe, this would now appear as microwave radiation, like that which would be emitted by a body at a temperature of around 5 degrees above absolute zero. Perhaps because the astronomers who did this work were rather fringe figures, and perhaps because the predicted radiation would not have been easy to detect, no one tried to look for it.

In 1964, two radioastronomers, Robert Wilson and Arno Penzias, were working with a radio antenna at Bell Laboratories in New Jersey. They were studying a supernova remnant, trying to work out the temperature of the radiation it was emitting. However, they kept getting other radio noise, which they did not want. They first noticed that this was constant whatever direction they looked in. So it could not be coming from a source on Earth. They thought it might be a fault in the antenna itself, perhaps caused by the pigeons nesting in it. So they tried to get rid of the pigeons (and their droppings), but the radiation persisted. It was microwave radiation corresponding to a temperature of around 3 K. They were reluctant to publish because this finding did not make any sense to them.

Meanwhile, unknown to Penzias and Wilson, two Russian astronomers had recently published a paper saying that something must be wrong with the big bang theory because the predicted background radiation would surely have been observed by antennae such as the one at Bell Laboratories. A theoretical group at Princeton University had also predicted the existence of radiation from the big bang – but were not aware that there was an antenna just down the road that was the best one

Figure 16.15

Dishes of the very large array radio telescope in New Mexico. The data from the 27 dishes are combined by computer to form a single radio image. In this way the array acts like one giant radio dish. The telescope is used to study radio-frequency radiation – rather than visible light – from other celestial bodies. Radioastronomy dates from 1932, when Karl Jansky discovered that many objects in the universe emit radio signals.

in the world for detecting it. So they started to build one of their own. In January 1965, Penzias happened to mention the radiation he and Wilson were observing in a phone call with another astronomer colleague, who happened to have heard from another friend of a talk by one of the Princeton group. The penny dropped – and in May 1965 Penzias and Wilson published a paper on their results, alongside one by the Princeton group which interpreted the findings. They had found the evidence for the big bang.

Penzias and Wilson were awarded the Nobel Prize for Physics in 1978 – as a result of finding something that they weren't looking for, and not realising what it was when they had found it.

Question

13 Before Penzias and Wilson's work, what reasons might astronomers have had for doubting the big bang theory? How did Penzias and Wilson's work provide evidence to support the theory?

The life history of stars

Kirchhoff's work on spectra enabled him to conclude that the Sun consisted mainly of hydrogen and helium, with other heavier elements in its atmosphere. In the mid-20th century, scientists used their growing understanding of radioactivity and nuclear reactions to work out how stars generate their energy. It comes from nuclear fusion reactions in the star's core, in which hydrogen atoms are fused together to form helium. *Ref to page 210*

Stars form as huge clouds of hydrogen and other gases are pulled together by their gravitational attraction. As they collapse together, the temperature rises enormously, starting the nuclear fusion reactions off. The total mass determines the temperature it reaches – and where the star appears on the 'main sequence' (Figure 16.7).

When the hydrogen is used up, the core of the star collapses, but its atmosphere expands and the star becomes a *red giant*. Other nuclear fusion reactions involving helium keep it going until the helium is all used up, and it becomes a *white dwarf* star. This will happen to the Sun, but it will not reach the red giant stage for another 5000 million years.

A white dwarf may explode (a supernova) or collapse further as its fuel runs out to become a tiny, very dense *neutron star*. (A matchbox full of matter from a neutron star would have the mass of an oil tanker.) It may even become so small and dense that the gravitational pull at its surface is too large even to allow light to escape. (Remember that Einstein's General Theory of Relativity predicts that light is bent by a gravitational field.) It has become a *black hole*.

An expanding universe: what does it mean?

If the universe is expanding, what is it expanding into? What lies beyond the universe? Will it go on expanding for ever? In what direction is the centre, where the big bang occurred? These are all questions that people ask. Perhaps you have asked them yourself. Some of them, however, are a misunderstanding of what the expansion of the universe means. It does not mean that the galaxies are expanding *into space*. It is *space itself* which is expanding. The only way to imagine this is through an analogy.

Imagine that we were two-dimensional creatures, living on the surface of a large sphere. We only know the two dimensions of the sphere's surface; we know nothing about up or down. So we don't recognise that our space is curved. In fact it has no end, and there is nothing outside or beyond it. Keep going in the same direction and you will eventually come back to where you started. This is a finite, but unbounded universe in two dimensions.

Now, instead of a sphere of fixed size, imagine that we are on the surface of a balloon which is being steadily blown up. The distance between any two points is getting bigger all the time. Then translate that image into three dimensions (if you can). This is how the universe is expanding.

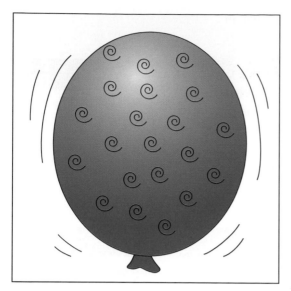

An expanding universe in two dimensions. The galaxies are on the surface of a balloon which is being steadily inflated. Measured along the curved surface of the balloon, every galaxy is moving further away from all the others. For any point, the galaxies furthest away are moving fastest.

Discussion points

How is our knowledge of the universe different from that of people 500 years ago? What are the most significant differences in the things we know about it? Do you think this knowledge affects the way we live?

Imagine that you read in a newspaper article that 'scientists have detected radio signals from a galaxy 256 million light years away'. What might make you suspicious about the trustworthiness of this article?

Every point is getting further from every other point. That is why it seems to us that the distant galaxies are all moving away from us – as though we were at the centre. We are not. They would seem to be moving away from you wherever you looked at it from. It is as though they are on the surface of that inflating balloon (Figure 16.16).

Will the expansion go on for ever? Cosmologists are not sure. It depends on the total mass of the universe. If it is large enough, the gravitational attraction will eventually slow the expansion and perhaps reverse it. This has led to efforts to estimate the total mass of the universe. There is evidence, from studies of the rotation of galaxies, that these contain large amounts of 'dark matter', which emits no electromagnetic radiation and so is invisible. Perhaps as much as 90% of these galaxies' mass is dark matter. If so this would greatly increase the total mass of the universe – and might mean that it would not expand indefinitely. But these questions are not yet resolved. Cosmologists hold different views.

So what is the direction towards the centre, where the big bang occurred? The balloon analogy helps us again here. Imagine that the balloon has been blown up from a very tiny starting size. Although we now see all other points on the surface moving away from us, there is no single direction that is back towards the start point. Every direction looks back towards it. That is why the background radiation from the big bang, that Penzias and Wilson observed, is uniform, whatever direction you look in.

Do we live in a special place in the universe? The answer which modern cosmology gives is no. Perhaps the question might be better asked as 'are we special?' In one way, we certainly are. For it is rather remarkable that a species has evolved on one small planet, orbiting a fairly ordinary sun, located in an ordinary position in a typical spiral galaxy, which has been able to work out where it is.

Review Questions

14 Make a time line to show the main stages in the development of our understanding of the structure, size and age of the universe.

15 Explain how Cepheid variable stars have played an important role in helping us measure distances to other objects in the universe.

16 Building large telescopes on Earth and putting the Hubble space telescope into orbit cost millions of pounds. List the arguments for and against spending large sums of money on astronomical research. Do you think this area of scientific work should be supported by governments?

17 'The Sun is an average-sized star, towards the outer edge of a medium-sized galaxy. The diameter of this galaxy is around 100 000 light years. The next nearest galaxy is 2.2 millions light years away.'

How should you regard these statements? Are they facts, or estimates, or hypotheses, or theories, or what? How good is the evidence supporting them?

Chapter 17

Coursework guidance

This chapter is for students who are following a course leading to the GCE AS in *Science for Public Understanding*. If you are entering for this qualification you have to submit two pieces of coursework. One is a *Study of a topical scientific issue* and the other is a *Critical account of scientific reading*. You should refer to the specification for details of the assessment criteria and the marking scheme. You can download the specification from the qualifications section of the AQA web site (http://www.aqa.org.uk).

You can do your two pieces of coursework in either order, or even together. In this chapter the advice for your *Study of a topical scientific issue* comes first simply because it is listed first in the specification.

The *Study of a topical scientific issue* gives you the chance to find out more about an issue which is not covered elsewhere in the course – maybe it has hit the news since this book was written, or is of more particular local or personal interest. You are asked to seek out the necessary information and write a report to show that you have developed your understanding of the topic and the science involved.

The *Critical account of scientific reading* gives you the opportunity to read some popular science writing. You are asked to write a report which shows that you have developed an understanding of the scientific ideas involved, can express your personal response to the issues and assess its appropriateness for the intended readership.

> **TIP**
>
> The advice from students who have completed this course already is 'plan ahead'. Give your work the time it needs and work steadily and evenly over the weeks you are given. Your deadline will come all too quickly, and the amount of other coursework you have to do in all your subjects builds up even more quickly.
>
> Don't be the one who says, 'I wish I hadn't left it till...'.

The Study of a Topical Scientific Issue

Choosing your topic and title

Choose your topic carefully – it may be the single most important thing you do. If you get this right the rest will follow.

Choose a topic that:

- interests you;
- allows you to do everything needed to meet the assessment criteria;
- is well documented.

It is a good idea to investigate an issue which you find interesting and will enjoy, but you may not produce your best work if you stick with something you already know a great deal about. If you are already familiar with most of the arguments and have made up your mind on the issues it can be more tricky to gain marks for seeking out and organising information. On some issues there may be heated debate, but that debate may not really involve much science. If you

> **TIP**
>
> Avoid a topic if there is little available scientific explanation or evidence – even if it looks interesting. For example, topics involving the supernatural fall into this category. They are not impossible, but a student who chose 'Spontaneous human combustion' as an issue found that she spent many more hours than she had planned tracking down rare scientific data, with a disappointing result.

already have strong opinions, there is a danger that you will leave out the science behind the issue or fail to present opposing points of view. For example, students have found it hard to gain high marks for studies of social issues such as the right to abortion.

Once you have a topic and have agreed it with your supervisor, create a working title. It is a good idea to make the working title a question. Then in the rest of your work you set about answering it and use it to judge the relevance of any information you come across.

Suppose you choose 'diabetes' as your topic, your working title might be 'What is the recommended treatment for diabetes and is it available to everyone on the NHS?'. The question gives an immediate direction and point to your enquiry. As your work develops and you gain more understanding of the issue, you can extend or change your working title.

Planning your research

With your final deadline in mind, plan your research so that you have plenty of time both for considering the results and for writing your report.

Divide your topic into sections for research based on your working title. Work out what questions you are going to try to answer in each section. For each question you are working on, consider the best places to search for answers. For example, you might best find out what diabetes is from a medical text or from the internet, but finding out how it feels to receive certain treatments might require an interview with a diabetic. Be prepared to follow up new questions your research raises.

Figure 17.1 lists a number of possible sources where you can begin to search for information. Few will use all of the suggestions – some topics will rely more easily on some sources of information than others.

Organising your work

Make sure you know your final deadline and any important intermediate dates. Your supervisor may want you to present a summary of your research at an early stage or may specify a last date on which you can hand in a draft for advice. You may also have subsidiary deadlines, for example, a presentation of your work for 'Key skills evidence' to include in your planning. Create a time line for your work.

Once you have begun your research you may find that you are led to visit places and to read articles you had not considered before. This is all part of the excitement of research. It is similar to the job of a detective: tracking down information, asking questions, keeping notes and going on to find out more. So always be ready to change your plan as you go along but do start with a plan. Consider the research needed for each part of your report, and check opening times of places like libraries and museums. You do not want to be surprised by a holiday closure just when you intended to work.

As you research, keep notes. These should contain a reference to all the things you want to write in your final report. Keep a file of any

Key terms

In this chapter the word **research** mainly refers to the type of library, text or internet study which you will undertake for your coursework. This is similar to the work of a science journalist 'researching' a story for the press, radio or TV. This is distinctly different from the **scientific research** which has been described in Chapters 1 to 16.

TIP

In order to gain marks, your report must show a variety of research so that you can demonstrate discrimination and assess reliability of different sources.

TIP

Discuss progress on your study of an issue regularly with your supervisor (and Key Skills Tutor, if appropriate). It may be they will suggest other members of staff who can be helpful for particular topics or in particular situations.

Internet
There is a huge amount of authoritative information quickly and easily available on the Internet through web-sites and, for many issues, newsgroups and lists. If you have access to the Internet, it is certainly an excellent start to finding information and to giving you an idea what is available. The resources section of the web site for this AS course (www.nuffieldfoundation.org/spu) will get you started with links to many relevant web sites related to topics in the course. If you have chosen a topic not included in the course use search engines to locate appropriate data.

School/college library
It is likely that your own school or college library will stock books and journals (such as *New Scientist*) which relate to the course. Ask your librarians what sources they would suggest.

Local public library
You are probably already a member of your local library. Your local library will have a different range of books from that in the school or college library. A local library can easily order books for you from other libraries. Many also now have access to useful online databases. Discuss your research needs with the librarians. They will usually be very pleased to help you.

Specialist library
These are libraries which specialise in particular topics. Some will also be public libraries, others may be linked to a manufacturer or association. They often have restricted access or will be geographically inconvenient for you, but some will deal with enquiries by phone. Others will help if approached through public librarians

TV/Video
Make notes from programmes which are about your topic including the national and international news. Keep a note of the details of the programme, channel, date, so that you can give full references if you decide to include the information in your report. Check ahead in the listings in magazines or in newspapers for programmes relevant to your topic so that you do not miss them.

Newspapers
You will find issues to do with the public understanding of science in newspapers every day. Weekend newspapers, in particular, often carry excellent features and are very well presented. Keep an eye out for items related to your issue. Many broadsheet newspapers now have archives which are accessible through the Internet or available in schools and colleges on CD-ROM.

Magazines and journals
Thousands of different periodicals are printed each week, each month, each quarter. Useful articles can be found, particularly in specialist magazines related to your topic. Magazines like *New Scientist* and *Scientific American* will certainly include articles relevant to many coursework issues. Use the index to search through back copies (most libraries keep them and some are now available online or on CD-ROM).

Museums and exhibitions
Many topics may relate to exhibits in a local or national museum. Visit the museum web site so that you can plan your trip and make good use of your time. Ask museum staff for help. They will be able to answer questions and to direct you. If travel is difficult, many museums will respond to specific queries sent by letter.

Interviews
Consider interviewing an 'expert' in the field of your topic or someone who holds strong views/opinions/ideas. Always have a detailed list of questions ready and be ready to take notes or to tape-record the interview.

Surveys
Some topics lend themselves to surveys of knowledge, attitudes or opinion. Take advice when devising your questionnaire. Before you start you must tackle issues such as how to select the sample for the survey, the sample size and how you intend to analyse and interpret your results. Consider a pilot survey with a small sample to identify any problems with your initial ideas.

Letters
You may seek information by letter from organisations with an interest in your topic. Always be quite specific about what you are asking the reader of your letter. Before you write check that the information is not available on a web site. A request for 'anything to do with the environment' either will be ignored or will encourage the reader to send unwanted information. Only well-written letters are likely to get a useful reply and a stamped self-addressed envelope will help. Always write in a polite tone   never demand information. Contact lists and addresses can be found in a number of places: telephone directories, Thompson Local Directories, in books, and in magazines. When writing your letter say: who you are; what you are doing; why you have chosen to write to them and, as clearly as possible, what you hope to get in response. Don't rely on getting a reply. It may take a long time before you get a response, particularly from busy voluntary organisations, and they may not have the time or money to reply at all. If there are a number of possible places you could write to, then send all the letters at once and give yourself a good chance of getting at least some reply quickly.

Telephone
If you simply want an advertised leaflet or, perhaps, to find out whether it is worth writing or who to write to, then you may prefer to telephone. Prepare exactly what you are asking for and be sure to give your name and address. Keep a note of the date of your call and who you spoke to.

Family and friends
Discuss your ideas with people you meet outside school or college. You may get suggestions as to how and where your research could develop. Or you may discover 'experts' among these people – their hobbies, their jobs, their personal history could be relevant.

TIP

Photography: your topic might lend itself to photography either to record a local 'before' and 'after' situation or as evidence of your being 'there'. Keep a record of your photographs, including where they were taken, when, and what they show.

Tape recording: you may want to record something other than an interview, for example, two people debating or discussing your topic, or evidence of results from a practical investigation such as noise levels. When recording other people, ask permission first.

Experiments/tests: as part of your research consider an experiment or test to develop your ideas. For example, you might make a quick check of pollution levels in a local river. These do not have to be complicated. Your supervisor may be able to help and loan equipment.

TIP

Don't be the student who is saying to their supervisor on the day that coursework is due in, 'I know I've got to put the name of the book I used on the form but I've forgotten – it was something like *Genetics for Now* and big and red and on the top shelf in the library, but it's not there now.'

leaflets and smaller bits of paper such as news cuttings, compliments slips and photographs. Organise your notes so that you can easily refer to relevant records when writing. For example, you could number all the pages in a notebook and have a contents page at the beginning, or use it as a diary so that it follows a time order. As a minimum you should keep an up-to-date record of:

- all the people/organisations you have contacted, together with dates, addresses, telephone numbers, and so on, as well as the dates of any replies you may have received;
- all the books, magazines, journals, web sites, and so on that you have consulted – keeping full records of sources as you go along will save you hours of time when completing the list of references in your report; *Ref to figure 17.5 on page 232*
- any other sources you have used.

Some people find it helpful to note references on index cards or on computer file. This aids sorting later.

Key Skills

If you are taking a qualification in Key Skills as part of your advanced course then the process of completing your coursework in *Science for Public Understanding* provides excellent opportunities to gather evidence for your portfolio. If you want to benefit from this opportunity you should talk with your Key Skills tutor at an early stage so that any necessary collection of evidence, or log book requirements, are in your planning from the beginning. The coursework tasks together contribute most readily to portfolio evidence for *Communication, Improving own learning and performance, and Problem solving*. Possible contributions to *Application of number and Information technology* depend on your choice of issue, how you intend to deal with any numerical data you collect during your research, and how you plan to present your work.

You may find that your Key Skills Tutor suggests additional tasks which are not a requirement of the Advanced Subsidiary assessment. For example, you might make an oral presentation of some of your research findings to others in your group for *Communication*. You should find that such additional tasks add depth and quality to your work without disruption, as long as you plan them from the start.

Evaluating information about science in the media

This section suggests ways of responding critically and reflectively to articles in newspapers or magazines, or when listening to reports on television or radio. You do not have to adopt these approaches in your study of a topical issue but you may find them helpful.

It is always a good idea to separate the 'facts of the matter' from the conclusions and explanations that people put forward.

Looking critically at data

Although data are usually more reliable than explanations and predictions, you should not just assume that the 'facts' are right. If an article or media report presents some data, then Figure 17.2 shows some questions you might ask about this information.

- Is the data reliable? Have measurements been repeated? If so, do the results agree?

- Has the data been checked by anyone else?

- Is the data reasonably easy to obtain, or is it a tricky thing to observe or measure?

- Do different scientists involved agree on the data?

- If the data comes from a sample of some kind, is this sample big enough? Is it a suitable sample to have chosen?

- Is it real measured data, or has it been calculated from measurements of something related? Or from a computer simulation?

Figure 17.2

Questions to ask about data in a report.

Looking critically at explanations

Even if the facts are generally agreed, it is still perfectly reasonable to question the claims in the article about explanations and predictions. Figure 17.3 suggests some useful questions to ask.

- Is the explanation based on generally accepted scientific ideas?

- Are there other data and evidence that support this explanation?

- Is this the only explanation, or are there other competing explanations?

- Are the predictions about situations similar to those already investigated? Is it reasonable to extend the results to these new situations?

Figure 17.3

Questions to ask about explanations and predictions.

Who to trust

Often articles quote some of the scientists involved. Radio and television reports may include interviews with these scientists. How do you decide whether to accept these claims? Indeed, many reports present differing views, from two or more scientists. How then do you decide whom to agree with? There is no infallible 'recipe' for making these decisions – but Figure 17.4 highlights some of the things you might bear in mind as you reflect on news stories about scientific issues. You may not always feel you have the background knowledge to 'score' for all these aspects, so concentrate on those where you feel you can make a judgement.

Try using Figure 17.4 to assess an expert's comments in newspaper or magazine articles related to your study. If you award a score (from 5–1) for each aspect, and add these, does it help you to decide which views to take more seriously? What are the weaknesses of a 'scoring system' like this?

Thinking along these lines will help you when it comes to writing the discussion section of your report and evaluating your findings.

Figure 17.4 Assessing how far to trust an 'expert's' views.

SCORE	ASPECT				
	Theoretical ideas involved	**Nature of the data**	**Status of the scientist**	**The scientist's institution**	**Personal affiliation**
5	Core science – agreed by all	Reliable and agreed experimental or observational data	A recognised authority in this field	A famous university or scientific research institute, or a major company	Works for an official regulatory body with responsibility for this area
4	Agreed by many, but still contested by a few	Experimental or observational data that is challenged by some	A professional scientist working in the area – but not a top name	A known, but less prestigious institution or company	Has no direct personal or professional interest in the issue
3	There are several competing explanations in this field — and this is one of them	Data that is agreed to be sketchy and uncertain	A respectable scientist but whose expertise is in a different field	An institution or company with a more doubtful reputation	Has been involved in these issues for a time and is known to hold a particular view
2	A new field in which there is no agreed theory as yet.	Data calculated from computer models, or projected from other data	A relatively junior scientist with no established reputation	An institution or company which few people have heard of	Has known views or contacts that might bias views
1	A fringe theory accepted only by the author and his/her friends	Data little more than an educated guess	A known maverick (or crank)	Not employed in an academic or scientific research institution	Works for a company with a direct interest in the issues

TIP

You can handwrite your final report or use a word processor. However, there is no difference in marks unless your handwriting is actually unreadable. If you are handwriting, use one side of the paper and start each section on a clean page – it will then be easier to substitute an amended page if you need to. Use double spacing for word-processed or handwritten reports as it is so much easier to read. Always number your pages as the last thing you do.

It is good practice to set your report out like a short book rather than as a long essay. This means that you will have a title-page, contents and so on.

Writing your report

You have to write a report on your topical scientific issue which includes:

- a **title**
- an **abstract** of the report (50–150 words)
- the **main points** of the report with the findings
- a **discussion** of the findings reviewing arguments on both sides of the issue leading to
- your **personal conclusions** drawn from the evidence
- a list of **references and sources** used in the study.

Remember your audience. This is a course in science for public understanding so you should write a report which explains your understanding of the issue to a 'general reader'. Specifically, though, you are writing for your supervisor who will read and check your work, and an examiner appointed by the Awarding Body who will also read your report.

Plan the structure of your report so that you know how you are going to use your material and roughly how much you will need to write for each section. The total length should be 1500–2000 words – no more.

Your final report may have even fewer words if you are going to present a lot of your information in tables, graphs, flow-charts or labelled diagrams. Illustrations can help a reader by breaking up the text but in your report there is a very limited place for pictures that are just there to 'make it look pretty'.

Plan how and where your report is going to do justice to all your research and thinking. As you are writing, make sure each sentence and paragraph is contributing to your marks, not only by being relevant to your argument and either presenting detailed knowledge or contributing to your overall evaluation, but also by making the stages in your work clear. For example, in your introduction to your report, *Diabetes care for different ethnic groups*, you might write:

'There are two main types of diabetes which are defined as...'.

This is fine and would begin to contribute to your marks for 'Content'. However, if you read several different leaflets and book chapters on diabetes before finding a clear definition that you found useful, you can write:

'Among a number of publications, the best definition of the two types of diabetes I came across was ...'.

In this way you are showing the examiner the breadth of your research as well. You are gaining credit for finding material and making a sensible decision to leave it out. It is very important to ensure that evidence of all your hard work appears in the report.

When you are writing your report, make it your own. You are expected to be using sources of information – indeed you are credited with marks for finding those sources and picking appropriate parts to add to your arguments. It is virtually impossible to get good marks if you do not use information from somewhere else. You must, however, avoid any possibility of being judged guilty of plagiarism, which is representing someone else's work as your own.

Plagiarism

'plagiarise, to steal from the writings or ideas of another'.

Chambers English Dictionary,1990.

When you hand in your coursework you sign a 'Candidate record form' for the Awarding Body. You should have included all your references and sources as appendices to each of your pieces of coursework. If you left anything out, there is the space to add it here. The form reminds you that, 'To present material copied from books or other sources without acknowledgement will be regarded as deliberate deception.' You also sign to say you have read a 'Notice to Candidate' which says: 'The work you submit for assessment must be your own. If you copy from someone else or allow another candidate to copy from you, or if you cheat in any other way, you may be disqualified...'.
It is not just in examinations, but also in academic work generally, that the crime of plagiarism is taken very seriously. Plagiarism is not just copying directly and not acknowledging that you have done so; it is also following someone else's idea or argument too closely.

TIP

Do not assume your reader has any previous knowledge of the subject you are writing about. This will encourage you to write clearly. Imagine someone else in your class is going to read it. Will they understand your account?

Consider including a glossary of any special terms as an appendix.

TIP

A short, well-organised and carefully argued report is better than a longwinded rambling account of a topic.

Do not add leaflets, booklets, print-outs or photocopied documents to your final report. There are no marks for bulk or excessive length or for work done by other people.

TIP

Do not rely on your supervisor knowing you did some difficult research work; anyone must be able to see, by reading your report, that you did that work.

Here are three ways to avoid plagiarism:

- When you are ready to write the first draft of each section of your report, look through any books, notes and other material you have accumulated. Then put them out of sight and write that draft from your head. Do not worry if you cannot remember the odd detail or want to put in a quote, just carry on. When you have finished the section, go back to your books and notes for quotations and missing details which you can now add to give more substance to your work.

- If you copy the actual words from your source, put the words in quotation marks ('...') and give a reference (see Figure 17.5). Similarly, if you copy pictures or diagrams you must give a reference. If you slightly alter a diagram to make your own point, then write that it is 'adapted from ...' and give the reference.

- If you outline a particular idea based closely on one source, or use specific information from a source, then give the reference immediately. If you have used a number of ideas from several sources, then give the references at the end of the paragraph or at the end of the section.

If you work in this way you should find that the pattern and thread in your writing is your own, and that you are interweaving ideas and information from different sources (both of which get you more marks). You will be giving clear references which are indications of your research and selection of material (which gets you marks). You then avoid risk being accused of plagiarism.

Be prepared to edit your writing – and do not be afraid to 'chop and change' it. Read and re-read it. Write and rewrite. Draft your sections and ask your supervisor to comment on them. Do not waste time presenting everything beautifully just to have your supervisor suggest a change at the top of the first page. Leave details like page numbering until the end. Your supervisor will probably prefer to see a second or third draft rather than your first rough notes but you may need to provide evidence of your developing work if you are building up a Key Skills portfolio.

Your title

Your working title is the question which guides your research. You can use the same title for your report but you need not do so. Often a short title is better but consider adding a subtitle. Choose a title which tells your readers what your report is about rather than a jokey title with no meaning. Remember that the content of your project is marked according to how relevant it is to your title and to what you say your report is about in your introduction – so make sure they match.

Your introduction

Your introduction should be a brief 'way in' to your report. It should arouse your readers' interest and outline what they are going to get out of reading your work. Include in your introduction a few sentences to describe why the topic is of importance to you or to the public. For example, you might explain a personal connection.

Consider starting with an opening sentence to grab your readers' attention and make them want to read more. For example:

> 'Harlequin stood on the platform, nerves tingling, as he watched for the pull of the lever which would send water pouring along the tubes behind him and launch him through the trap-door into an explosion of smoke above.'
>
> *[from a study of the use of new technology in the rebuilt Sadlers Wells Theatre]*

Other features of your introduction might be:

- A few sentences to explain the focus of your report.

  For example, if your report is called 'Alternative Medicine', you might explain that you actually have chapters on acupuncture, homeopathy and hypnotism, because you had discovered that these were the popular forms of alternative medicine and therefore good examples to use.

- A few sentences to describe the structure of your project so that your readers know where they are going.

  For example, in a study on animals in scientific experiments you might indicate that, after a chapter defining cruelty and pain, you have four sections giving examples of the use of animals in experiments, which you chose to show the possible range from necessary to unnecessary, and that you then have a section on the current law, before you give your opinion as your final conclusion.

- A final sentence which states the question that the rest of your report answers.

  When you have written the rest of your report, return to your introduction to check that it accurately reflects what you've written. You might even consider leaving the final drafting of the introduction until after you have finished the bulk of the report.

The main body of your report

The main points you have selected from your research will make up the main body of your report. You should already have created, on separate paper, an outline of what you are going to write and in what order. Writing over a thousand words can be a daunting task but it will be more manageable if you split your work up into smaller parts. Organise your work into groups of related areas, and present these as sections with sub-headings. Do make sure that there is a clear logic to the order. Ideally each new point should be related to the previous one in some way. Each section should be titled and listed at the beginning of your project in a contents page.

Many pages of continuous text can be hard to follow and unattractive to the reader. Select relevant diagrams, graphs, labelled pictures or

> **TIP**
>
> Do not waste space telling your readers that you had to pick a topic for GCE (they know that), what else you thought of doing (they might wish you had done it instead), or how you wish you had done something else (may be true but too late now).

photographs with captions to summarise information. A well labelled diagram is not only a useful and attractive visual aid but it can also often save you a great deal of writing. However, do not include illustrations for their own sake. Make sure you refer to them in your writing and always give each a caption.

Your discussion

In this section of your report you review the work you have carried out and justify your findings. This section might be quite short and concise, but it is essential to demonstrate that you really have understood and thought about the issue you have studied. It is possible, and sometimes preferable, to write some of this comment in your other sections as you go along, but it is generally a good idea to have a separate section as well. It keeps your thoughts clear and makes it easier for the examiner to give you the marks for 'evaluation'. This part gives your personal opinions, so it can be appropriate here to write in the first person ('I want to argue that'... rather than 'It is argued ...'). When expressing your own opinion, always justify your statements with reasons and examples. When analysing information, look at the subject impartially and give alternative solutions. Then base your conclusions on the facts you have discussed.

You should comment on the main points you have presented, for example:

- If different sources disagreed, how did you decide which to follow?

- If several references gave diagrams, how did you choose which one to adapt for your use; or have you used parts of one for clarity and another for completeness?

- If you have had to make judgement between different opinions, what questions have you asked yourself and how have you weighed the evidence?

- If you have come across any controversial question (or maybe your whole project is about a controversial question), how have you distinguished between facts and opinions?

- If you have paid more attention to some sources than others, was it because some were more recent or in a more prestigious journal – or because they matched your opinion?

- How did you decide what material to use and what to leave in?

Your conclusion

Your conclusion should include a clear statement of your own views on the issue you have been studying. It should leave your readers satisfied and absolutely clear about what you are saying.

Consider including:

- a reference to the opening sentence in your introduction which grabbed the attention of the reader – it rounds a piece off nicely to refer back to your introduction by giving your explanation or answer to the question or issue you started with;

- a summary (almost a list) of the most important points you have made and your main opinions or conclusions;

- a few sentences which indicate what further research needs doing.

Your conclusion should not include new information. You should already have given all the facts and your evaluation of them in earlier sections.

Your abstract

Write your abstract last. This may repeat phrases or sentences from elsewhere in your report, if appropriate. You need one or two short sentences to describe the range of points on your issue which you cover, followed by a further sentence or two to describe the main points and conclusions you have come to. The aim of the abstract is to tell readers what they are going to find in the report briefly and clearly. It goes in a separate section at the front of your report, but should always be written last since it must be accurate reflection of your final piece of work.

References and sources

You have to list references in a separate section at the end of your project. Number your references or list them in alphabetical order – so this is a job you should leave till last. Remember to name anyone who has helped you. If you have discussed your project with anyone or been given any advice, acknowledge this help. The only exception to this rule is that you do not have to credit your supervisor.

It is very important to have kept a clear record of your sources as you went along so that you can make your list of references accurately and quickly (see Figure 17.5 on page 232).

You must list:

- any books, periodicals or other media you have quoted in your report;

- any books, periodicals or other media from which you obtained background information;

- any people, other than your course supervisor, who helped you;

- any computer software packages you used.

> **TIP**
>
> Before you make your finished draft and hand in your report, you should make some final checks. Make sure that your report contains everything asked for in the specification for the course. Read through your final report one more time, making sure as you go that nothing is missing, and make any necessary last additions or amendments.

Books and periodicals

There are several different ways of citing references. For example, you might list your references in the order they occur in your report, number them and give the numbers in your report as you go along. It doesn't matter which style you use, but if you are unsure try the following:

In your text put the name of the author and the date in brackets. For example:

'A common modern view of the way particular genes might affect behaviour has been summarised as: ...' (Lewontin 1998)

Then, at the end of your report, add a list of detailed references in alphabetical order by author.

For *books* give the author (last name then initials), date of publication, title of the book and the publisher. For example:

Strathern, P. (1997), *Hawking & Black Holes*, Arrow.

Notice that the title of the book is in italics. If you are using a word processor then follow this example. If you are handwriting then underline the title:

Strathern, P. (1997), Hawking & Black Holes, Arrow.

Often you will find that you are using an edited collection on a particular topic or reading about a piece of work in a textbook. For a piece in an book with an 'editor', give the author, date and title of the actual part you are using, and then give the full reference for the book. Use the abbreviation 'ed.' to show that the book has an editor rather than an author. For example:

Lewontin, R.C. (1998), 'Genes, Environment and Organisms' in Silver, R.B. (ed.), *Hidden Histories of Science*, Granta.

A similar format is used for articles in periodicals but you need to add the volume number or precise date. For example:

Casti, J. (1999), 'Firm Forecast' in *New Scientist*, 24 April 1999.

For something quoted from another source give the reference to the book you actually used. For example:

Faraday, M. quoted in Bragg, M. (1998), *On Giants' Shoulders*, Sceptre.

You may be using *pamphlets* or leaflets with information from companies or charities. Sometimes these do not have details of the author or date on them. Do the best you can, using the name of the organisation as the 'author'.

Other media

For references to other media (CD ROM, internet, TV programmes, and so on) you need to give information which is similar that used for referencing for books and periodicals. It should be possible for your readers to find the original source you used so that they can follow it up themselves in more detail if they are interested (and, of course, your supervisor and examiner may want to check on how you interpreted the source).

For *CD-ROM* or *video* give title, date and publisher. For example:

News in Action 5 (Science) (CD-ROM 2000), *Daily Telegraph*.

Genetic Engineering & Farm Animals (Video 1999), Compassion in World Farming Trust.

For *radio* and *TV* programmes give the title of the programme, the channel or station on which it was broadcast and the precise date of broadcast. For example:

Analysis: the MMR Vaccine (broadcast 9 December 1997), Radio 4.

For internet groups, lists or websites give the title of the site or newsgroup, the web address and the date or dates on which you accessed the information. For example:

Changing Transport data (internet 17 March 1999), www.youngforesight.org/student/database/transport.

For *people* who have helped, give their names, their jobs or relationship to your work, and a few words to say how they helped. For example:

Professor Jack Winchester, Professor of Oceanography at the Florida State University program in London, allowed me to interview him about the way Americans view environmental issues after I met him at a British Association event for young scientists, March 1999.

For *places* you have visited, give the name of the place and a few words to indicate what use you made of the visit. For example:

Worshipful Company of Goldsmiths, made notes and bought the postcards used for illustrations at exhibition on modern assay methods.

Account of Scientific Reading

The aim of this piece of coursework is to help you to discover the pleasure of reading good, popular science writing.

Choosing your reading

Your supervisor may direct you to some reading or give you some choice. If you are making your own choice, whether from everything available in your college library or from a more limited selection of recommendations, it is important to choose a piece which allows you to demonstrate the skills which gain marks in the final assessment. You are free to review any piece of popular science writing – fact or fiction. However, it needs to be a substantial piece of writing (a minimum of 30 pages or so of a typical book). You also need to bear in mind that you have to demonstrate that you understand the science in your reading, so you need to be careful to choose writing that has some science content – particularly if you are considering selecting a piece of science fiction.

If you are using a book as your source (as opposed to a short story, or lengthy article), you are effectively writing a review of the book, but you will use examples from the one section which you read in detail.

Since you are asked to make judgements about the appropriateness of the writing for its intended audience, you might want to consider looking at writing intended for a particular readership, such as 14–16 year-olds. You will be a good judge of the audience but also be able to 'stand back' with some more maturity.

The examiners particularly like to see you choose reading which has caught your eye because you have a special interest, but your supervisor has a list of sources from the Awarding Body which have been used before and which might give you some ideas.

Once you have a source for your reading, and have agreed it with your supervisor, the first steps are to find out a little more about the source and, of course, to read it.

Read and assess your text bearing in mind the intended readership. For books you can often get some indication about the author's intentions from the information on the cover, or in an introduction or preface.

For long articles from a periodical you need to judge who buys it. Your librarian may be able to help with some advice.

If you have access to the Internet you might try finding the source at an online bookstore. They often add reviews contributed by readers and you can gain a picture of the readership from them. (You might also gain other ideas – but remember that those reviews are not written with an examination in mind and note the warnings about plagiarism.)

 *Ref to pages 227–228*

> **TIP**
>
> The passage your choose to read must be taken from a source intended for the general public. A science textbook is not an appropriate choice.
>
> The specification from the Awarding Body does allow you to read two or three related essays or articles by different authors. This usually turns out to be more difficult than reading and reviewing one longer passage of reading from a single source by one author.

If you are starting with your *Critical account of scientific reading* and have begun with this section, do go back and look through the general advice on writing a report for the examiners in the previous section, as it is not repeated here.

Try reading some short book reviews in magazines or newspapers to see how experienced writers assess a book critically in a few words. Many book reviews are only about 250 words long.

TIP

If you are reviewing a few chapters of a book, it is a good idea to at least skim the introduction and conclusion of the book and the other chapter titles so that you are fair to the author. (The chapter you have said is impossible to follow might have been easier to understand if you'd read the chapter before, or it might fall into place if you read the following chapter too…) Of course there is nothing to stop you – and everything to gain – from reading the whole book anyway and just using examples from one section in your account.

Your written account

Your final account must be short, in the range 500–800 words. The best accounts really are short and succinct. You should aim to write no more than 800 words. You will not lose marks by exceeding the limit slightly but you will not gain marks by doing so.

The specification suggests that your account should include:

- a **precise reference** to the text you have read;
- a **summary** of the science ideas and explanations in the text, set in a wider context, if that is appropriate, and related to any social or moral issues;
- your **personal response** to the ideas and explanations in the text;
- a **critical discussion** of the effectiveness of the text for its purpose, its style and language and the values or attitudes it conveys to the reader.

There is no fixed style which you have to use for your account. You do not have to stick rigidly to the advice below and your supervisor may well have suggestions to make. Apart from the introduction and the conclusion, you do not have to write the components of your account in the order given below.

Reference

As in a book review, start with the details of what you have read including the title, author, edition, publisher, publication date and page numbers.

An introduction

Aim to open your account with two or three sentences which engage your reader and show why your chosen text is interesting and has something worthwhile to say. At the start you might set the passage in context and identify the intended readership.

Summary of the scientific explanations and ideas

You should then write a few paragraphs summarising the main science ideas or explanations in the passage you have read and any issues which arise. Remember that this is a short review and you are aiming to demonstrate that you have understood your text. So use your own words to explain the main ideas clearly, as if to a fellow student. If there is a lot of detailed science in the reading you have chosen, then you should select one or two ideas which stand out as good examples.

You may find that you need to look up some points in another text in order to understand your reading or to be able to make a comparison. If so, you should make this clear but do be careful not to start full-scale research into the issue in other sources. This is supposed to be a succinct review of one piece of reading. The place for wide research on an issue is in the other piece of coursework.

Add a few paragraphs commenting on any issues which arise from the topic of your text and comment on any social or moral issues which may arise.

Your personal response

You need to add a paragraph giving your personal response to the reading. Be honest (favourable or unfavourable) and give reasons and examples to illustrate the points you make. Remember to use appropriate language for a critical review. Do not lapse into very informal language just because you are stating your own opinions.

Critical discussion

You should finish with a paragraph or two in which you assess the effectiveness of the writing. It is perfectly appropriate to consider issues such as the type of illustration, the cost, and so on, as well as the actual style and language of the writing.

This is the place to include any comment on particular views or opinions the author gives, unless you have already made that a part of your personal response.

Conclude with a sentence summarising your view of the effectiveness of the text as an explanation of science ideas for its intended readership.

Handing in your coursework

When you are sure everything is as you want it, put the final package together. You may want to add an illustrated cover or title page. This is not essential, but it can give a final polish and create a good initial impression. Much more important is to make sure that all your pages are numbered and are firmly fixed to each another. It is perfectly sufficient to staple the pages firmly in a way that allows your supervisor and the examiner to turn the pages easily. If you want to make your presentation more professional, a *light* plastic binder can be useful, or you may have access to a thermal binder. If you use a slide binder, always staple the pages as well.

Warning

- Do not use ring binders.

- Do not put each page into a plastic wallet.

There are some formal details that you must include to comply with examination regulations. The main one is completing a Candidate Record Form and attaching it to the front of your report. It is important that you seek advice from your supervisor if you are unsure how to complete any of the details on that form. Your supervisor will also be able to tell you of any other procedures which you must follow in handing in examination coursework at your particular college or school.

18

Revision and exam preparation

The specification, the course and the examinations

The style of the examination papers reflects the style of the course. As you work through the course you study a series of teaching topics set out in section 9 of the specification (syllabus) from the Assessment and Qualifications Alliance (AQA). You can download the specification from the qualifications section of the AQA web site (http://www.aqa.org.uk).

As you study each topic you learn to apply your knowledge of scientific explanations (see section 13 of the specification). You also find out about the way that science works by considering a series of ideas about science (see section 12 of the specification).

The opening pages of each of the first 16 chapters in this book show you how the three strands of the course are interwoven: the issues, the science behind the issues and what a study of the issues tells you about science and society.

In a similar way the exam papers consist of a series of questions each of which starts with some information about a topic. You study the information and then answer questions which show that you can:

- make sense of the information and form your own views about the issue;

- explain or apply the science; and

- appreciate the significance of some general ideas about science and society.

So each exam question is like a mini-chapter of this book with information and questions to test your understanding and to promote the exploration of ideas. This means that you prepare for the examinations whenever you study a chapter in this book, answering the questions and discussing the issues.

Revision

As the time for examinations approaches you will need to revise. Revision in this course is, however, rather different from revision for other science examinations. The emphasis is much more on ideas and understanding and much less on being able to recall a large quantity of facts, patterns and theories.

Give yourself plenty of time for revision. You could make some revision notes as you complete a unit or in preparation for a mock exam. Leave time to go over past questions and to test yourself several times on each unit.

Figure 18.1 The three strands of the course in topic 10.4 of the specification: 'Fuels and the global environment'.

FUELS AND THE GLOBAL ENVIRONMENT

Ideas about science ('Data and explanations' and 'Causal links')	Teaching topics	Science explanations ('The particle model of chemical reactions' and 'The radiation model of action at a distance')
	The carbon cycle: natural reservoirs of carbon compounds, scale of fluxes between reservoirs.	All matter is made of elements. Elements can combine to form other substances called compounds. The properties of the compounds are completely different from those of the elements.
Scientists value measurements which are replicable.	Changes in the level of carbon dioxide in the atmosphere; evidence and possible causes.	A chemical reaction involves the rearrangement of atoms to form new and different substances.
We cannot be sure that a measurement gives the true value of the quantity being measured. We have more confidence in the average of several repeated measurement than in a single measurement and the variation in repeat measurements helps to identify the range within which the 'true' value probably lies.	Evidence of changes in average global temperature over time.	
There is a particular problem in 'proving a negative', that is, of obtaining convincing evidence that a factor does not cause a claimed effect.	The greenhouse effect and its significance for the Earth's climate.	Some objects can affect others at a distance by emitting radiation which travels through space. The hotter the object the shorter the wavelength of the radiation. The differences between the various types of electromagnetic radiation are due to their wavelengths. When radiation strikes another object it can be reflected, transmitted or absorbed. When radiation is absorbed it can cause heating.
Many questions of interest are not yet amenable to a full explanation in terms of a predictive theoretical model. Often the best we can do is to look for correlations between a particular factor and an outcome.	The link between atmospheric carbon dioxide levels and climate. Use of computer models to predict weather patterns; advantages and limitations of computer modelling.	

Revision notes need to include all the main points you have to recall but they also need to be short so that you can commit them to memory and go over them several times before the exam.

When you come to the exam you may have to use the science explanations and the ideas about science in connection with the issue where you learned them. But, you may also have to relate them to an unfamiliar issue or context which will be described in the question. Your notes need to be organised so that you are prepared to use your knowledge and understanding in this way, linking all three strands and knowing enough about each one to also use them in unfamiliar contexts.

There are many ways of organising effective notes for revision. Spider diagrams, tables, cards and linear summaries all have their devotees. Whichever style you find is right for you, you will have to be ready to make links between the three strands of the specification.

Revising the issues

Linking the strands of the course

Figure 18.2 shows one example of how you might write notes in table form for the issue of Genetic diseases (section 9.5 in the specification)

You should go through the whole specification making notes for each of the 13 issues in section 9. You should aim to get each issue on not more than two pages. If you write more than this, you are writing too much. Notes of this type are intended to summarise the key points. Writing them involves thinking quite hard about which *are* the key points, so making notes is itself one of the most useful revision activities. It makes you think about the material.

You might find it helpful to discuss what should go into the notes with someone else. Discussion is an excellent way to clarify ideas.

Giving examples related to the issues

There are places in the specification where you are expected to be able to give your own examples. You will spot these in section 9 by looking for 'e.g.s' in brackets. These are listed below. Make sure that you know what examples you will use in an exam question and make a summary of the key points:

9.1   Development of understanding of transmission of infectious disease

Study of one non-viral disease

Study of one viral disease

Development of preventive measures

9.2   A study of a claimed risk to health

9.3   Use of routine screening during pregnancy

9.5   Study of one genetic disease

10.5 A study of a possible risk to health due to low doses of radiation.

There are many other places where it helps if you can give examples even if this is not explicit in the specification. However you will usually find that in an examination the examples are given with the information at the start of the questions.

Formulating opinions

There are many issues covered in this course where there is no right answer. You are invited to make up your own mind. You will have discussed some of them in class. In the exam you will be asked to give your opinion. It is very important to remember that the marks are not for saying you approve or disapprove of something but for the justification of your position. You should not only mention why you have taken your position but acknowledge the counter-arguments and explain why you have rejected them. Prepare yourself for this by making notes on the points for and against some of the issues covered. This is probably best done in the form of two columns.

Figure 18.2 An example of notes on topic 6.5: 'Genetic disease'

IDEAS ABOUT SCIENCE	ISSUES	SCIENCE EXPLANATIONS
	Example - Cystic fibrosis (CF) Symptoms - mucous in lungs and digestive system Leading to risk of serious infection Life expectancy up to 30 years Treatment – no cure, physiotherapy, antibiotics relieve Gene therapy experimental, not effective yet	Inheritance of two sets of chromosomes, one from each parent Chromosomes carry genes Genes provide information for cell function Two copies in each cell, only one expressed the other is recessive CF caused by defect in single gene, recessive therefore only get disease if CF gene inherited from both parents Both parents must carry gene One in four chance of each of their children having disease One in two chance of their children being a carrier
	Ways of avoiding birth of children with an inherited disease	
Ethical principles Right to life Use of technology to prevent suffering	**1** Antenatal screening – testing a few cells taken from the fetus as it is growing and testing for defective gene Decision about abortion	
Right of individual parents to make choices which may be based on fundamental moral position Right of society to maintain health of population and control health care costs	**2** Genetic counselling – pre-conception parents tested for genes. Advised on risks. Allowed to make their own choice	
Rights of parents to choose and of society to constrain this choice	**3** Pre-implantation genetic diagnosis – IVF embryos tested and only healthy ones implanted (Potential for designer babies, with particular sex or traits. Access to technology by wealth.)	
Control and regulation of the application of scientific knowledge by society	**4** Gene therapy – use of genetic manipulation to insert an extra normal gene. Experimental work on somatic cells such as lungs of CF patient Safety issues of current trials Not allowed on human germ line cells at present although this has been used in animals Unknown long term implications of germ line therapy	Identification of specific gene Insertion of normal gene into a carrier, usually an inactivated virus Gene carrier administered to patient, random chance of the gene inserting itself and functioning

Examples of the issues you should consider in this way are:

- the priorities for prevention and cure in the health service;
- the use of animals in medical research;
- genetic testing and gene therapy;
- the validity of the claims made for a form of alternative medicine;
- genetically modified crops and farm animals;
- the best method of generating electricity in different parts of the world;
- the strategies which should be adopted to improve air quality;
- actions to reduce the levels of greenhouse gases;
- the risks to health of low-level ionising radiation;
- the risks to health of non-ionising radiations (e.g. from mobile phones) and electromagnetic fields (e.g. from high-voltage cables).

Revising the ideas about science

As you make your notes on the issues (as in Figure 18.2), check with the specification that you have included the relevant *ideas about science*. Also make sure that you relate these ideas to the issue you are studying. In the example in Figure 18.2 you need to learn the different ways of preventing the birth of children with genetic disease, but also relate this to general ideas about ethics and decision-making.

Section 12 of the specification sets out the *ideas about science* in general terms. This is because many of them apply very widely to a whole range of topics and issues. That is why they are useful ideas. You will find that you can apply these ideas to many topics and issues and not just the ones where you first met them.

You do not have to be able to recall the words in the specification. What matters are the ideas. The best way to show that you understand the ideas is to be able to give examples and explain their significance.

Once you have revised all the issues, go through the list of *ideas about science,* one by one, and check that you can find at least one example from the course for each idea.

Revising the science explanations

You will revise most of the explanations as you make your summary notes linked to specific issues (see Figure 18.2).

When you read the *science explanations* in section 13 of the specification it is important to understand that you do not need to know anything more than is explicitly stated. Many of the things you had to be able to recall and do in other science examinations are not required in this course. For example, you do not have to be able to write symbol equations for chemical reactions, explain the laws of reflection or refraction or draw and label detailed diagrams of the eye. Nor do you have to be able to carry out many different types of calculation.

A useful way of checking that you understand the words in the specification is to translate each of the key points from text into a diagram. You could draw a labelled diagram of the model of an atom or of the solar system or the key features of a living cell. You can review your diagrams quickly during last-minute revision. The work involved in planning the diagram will help you understand and learn the explanation.

Some examination questions ask you to explain a scientific concept to a member of the public, in other words making it easy to understand, not using specialised technical language. Practise this by talking to someone in your family about the explanations you meet in this course.

Hints for examination success

The format of the examination papers

There is one examination paper for each of the two modules. There is no choice of questions. The format for both papers is the same:

- total time   1 hour 15 minutes
- total marks   60
- number of questions   normally 4 or 5

Between 6 and 8 marks on each paper are awarded for the quality of written communication.

In the examination room

Reading the question

Many questions start with information which may be about a familiar or an unfamiliar issue. The questions then ask you to comment on the information, often bringing in other knowledge from the course. Start by reading this information carefully, underlining key words. Read the questions and then go back to the information to see what points there will help with the answer.

Making sense of the data

In many questions the initial information is in the form of data, often a table or graph. Take time to make sure you have fully understood the information.

Look very carefully at the headings. What is the graph about? Look very carefully at the units being used. Do the numbers represent absolute values or relative values? For example the total number of deaths in the UK from AIDS is different from the number of deaths from AIDS per 10 000 population. This is very important when you are making comparisons between different countries. Percentages are always relative values, but relative to what? For example, natural gas as a percentage of all fuel used is different from natural gas as a percentage of fuel used to generate electricity.

In many situations it is the rate of change rather than the absolute value that is important. This is indicated by the gradient (or slope) if the results are plotted on a suitable graph.

Experimental results are often presented as scatter diagrams. The important point is to see whether there appears to be an overall trend despite wide variation. You can often sketch a line showing the trend. This is helpful in answering questions about the data.

If you are discussing the significance of data, remember to look at the sample size and the use of controls. You might also need to know who collected it.

Checking the instructions

You do not have a lot of time in an exam so don't waste it by writing things that are not required. Follow the instructions. Note how many marks are allocated to learn how much detail is needed.

Every year too many well-prepared candidates fail to score as many marks as they should because they do not answer the question set by the examiners. Examiners try very hard to set questions which are clear to the candidates. Even so, under examination conditions it is all too easy to rush into writing an answer before checking carefully the meaning of the question.

A useful first step is to highlight the words in the question which give instructions.

Name   Means just that. 'Name two of the gases emitted from motor vehicles engines which cause air pollution' requires two names, for example: nitrogen oxide, carbon monoxide.

Describe   This is also about factual recall but requires some detail of the process or structure.

Outline   Has a similar meaning to describe.

Sketch   Means draw a diagram or rough graph. Do not forget to label the axes and give a title to the graph whether this is asked for or not.

Explain   This requires a more detailed answer. You may be expected to give a scientific explanation from the specification, or the question may want you to describe your interpretation of data or your understanding of a scientific development. The mark scheme will suggest how many points you need to make.

Suggest   This term is widely used in exam questions. It means that you have to think about your answer and apply what you know to an unfamiliar context. The expected answer is not given in the data nor can you simply recall it from the specification. A wide range of suggestions is usually accepted. Any reasonable suggestion will gain marks.

Estimate   Requires a numerical answer, but not an exact one. You may need to do a calculation, using values which are also estimates.

Calculate Means that you must show the steps in your working, and make clear what each step means. At the end of a calculation always check that your answer is reasonable and that you have stated any units. For example, if you come up with an answer for a number of people close to the total world population (6 billion) it is probably wrong. Did you use the correct number of zeros throughout the calculation?

Answering the question

Make sure that you answer the question rather than writing about anything you know. For example, if a question asks about harm to the environment you will not get marks by writing about harm to human health. Avoid general statements which could apply to any issue or are just a rewrite of the question. For example, you will not get marks for statements like 'it is harmful'. You need to explain by saying 'the radiation may cause mutations which lead to cancer'.

Writing a coherent story or argument

The examiners assess the quality of your written communication in longer answers, usually those with five or more marks. It is perfectly acceptable to answer shorter questions with one-word answers, diagrams, flow charts or lists of points. Do not, however, use abbreviations that are not widely accepted in English.

Where the examiners are assessing communication they hope to read a passage with the idea expressed in grammatical sentences and organised into a sensible structure. For example, if you are debating whether genetically modified soya beans should be grown, they would expect all the arguments in favour to be separate from all the arguments against. Your own point of view should also be clearly stated if it is asked for in the question.

Index

abortion 80–3

absolute zero 218

acupuncture 69, 70–4

Adams, John Couch 204–5

agriculture

    genetic engineering 87, 89–95

    global warming impact 164

AIDS 8–10, 40

air quality 125–6, 143–54

    controls 151–4

    and health 149–50

    monitoring 146–9

ALARA principle, radiation exposure
    177

allergies 65, 67

alternatives in medicine 62–74

amniocentesis testing 79–80

Andromeda galaxy 214–15

angina 54–5

animals, in scientific experiments 26, 32–7, 96

antenatal testing, genetic diseases 78–82

anthrax 7, 18–20

antibiotics 28–9, 30–1, 39–40

    resistant bacteria 32, 112

    see also drugs

antibodies 14–15

Aristotle, philosopher 195

arithmetic increase concept, and evolution
    105

artificial selection, evolution 106

asthma 65, 149

astronomy 192–220

    ancient 194–6

autism, MMR vaccine 23–4

back pain 65, 72–4

bacteria 1–11

    and viruses 8–10

BCG vaccine 21

becquerel (Bq), radioactivity
    measurement unit 183

Beagle voyage 102–4

'big bang' cosmological theory 208, 217–19

biodiversity, global warming impact 164–5

biotechnology described 89

black holes, astronomy 219

Bq *see* becquerel

Brahe, Tycho, astronomer 198–9

Broad Street pump 5–6

cancer

    cell division 53

    ionising radiation 170–2, 181–5

    leukemia 183–5, 189

    lung 51–3, 176, 182–3

    radiation 168–72, 176, 181–5

    radiotherapy 167

    and smoking 51–3

Cannon, Annie Jump 211

carbon, radioactive 169

carbon cycle 156–7

carbon dioxide

    from fuels 144, 151

    global issues 156–9

carbon monoxide 145, 150

carcinogens 51–2, 150

carrier, of genetic disease 77

case-control studies, risk factors 47–8

catalytic converters 151

cells

    division 52–3

    basic units of living things 8

    white blood cells 14–15, 28

Cepheid variables, stars 211–16

CFCs (chlorofluorocarbons) 146

chemotherapy 27–9

    magic bullet concept 28–9, 31

Chernobyl 95, 135, 173–4

chiropractic 66, 67, 69

cholera 1, 4–6, 7, 18

cholesterol 57–9

chorionic villus testing 79–80

CHP *see* combined heat and power
    schemes

chromosomes 51, 52, 76–7, 88

climate changes 161–6

clinical trials, drugs 37–9, 65

cohort studies, risk factors 47

combined heat and power (CHP) schemes 131–2, 142

commonsense, science and 192

complementary medicine 62–74
    evidence of effectiveness 68–74

computer modelling, climate change 161–2

contamination, radioactive 172–3

control group, in scientific investigations 37–8, 47, 182–9

Copernicus, astronomer 196–8

correlation 51–61, 119, 179, 181, 186, 190

cosmology 212–20

cost benefit analyses
    air quality 143
    health screening 85–6

counselling, genetic 82

cowpox 16–17

creationism 110–1

crops see agriculture

cumulative doses, radiation 183

cystic fibrosis 77, 85–6

'dark matter', cosmological 220

Darwin, Charles 102–10

data, importance of 12, 25, 192, 208

and explanations 1, 12–13

Dawkins, Richard 113

decision-making involving science and technology 13, 23–4, 33–5, 37–9, 68–74, 78–86, 115, 179–191

developing countries
    diseases 53
    electricity generation 140–1
    energy use 115, 119, 121, 128

diet
    and health 14, 57–9
    heart disease 57–9

diseases
    developed/developing countries 53
    genetic 75–86
    germ theory 1–12, 19
    heart disease 53–61, 71
    incidence 182
    lung diseases 48–53, 176, 182–3
    medicines 26–42
    miasma theory 4–5
    preventing 13–25

spreading 10–11

DNA 88
    and evolution 114
    and genetic engineering 88–9
    ionising radiation effect 170, 181
    viral 9

dominant genes 77

Doppler effect 215–16

double-blind testing 37–8

Down's syndrome 76–7, 79, 80, 81

drugs
    anti-viral 41–2
    clinical trials 37–9, 65
    development 32–3
    ethical issues 26, 34–7
    and medicines 27
    molecular structure 32–3
    resistance 26, 32, 40–2, 112
    side-effects 26, 40, 65
    testing 26, 33–9

Earth, age of 109, 217

ecosystems, global warming impact 164–5

efficiency 118, 122–3, 131–2

Ehrlich, Paul 28–9

Einstein, Albert 216

electricity
    costs 137–41
    future 136–41
    generation 130–1, 136–41
    power lines 187–9
    secondary energy source 117, 129
    supplies 129–42

electromagnetic spectrum 158, 171, 186

ELF see extra low frequency radiation

elliptical orbits 200

energy
    conservation 118
    consumption 119–28
    demand 121
    efficiency 118, 122–3, 131–2
    and fuels 115–16
    future 126–8, 136–41
    in homes 122–3
    idea of 118
    primary sources 116, 117, 120–21, 130–1
    renewable sources 116, 129, 136–41
    reserves 121–2
    secondary sources 117, 129

sources 116–17, 120–2, 129–31, 136–41

TPES 120–1, 127–8

transfer 116–18

for transport 124–5

trends, global 120

engines, pollution and 125–6, 144–5

environmental issues

air 143–54

fuels 115–16, 121–8, 132–42, 144–54

genetic engineering 92–5, 97–8

radioactivity 173–6

enzymes

genetic engineering 89

penicillinase 32

epidemics 22

epidemiology 4–6, 48–9, 56–61, 181–91

ELF 188–9

heart disease 56–61

leukemia 183–5, 189

lung disease 49–53, 182–3

mobile phones 185–6

Eratosthenes, astronomer 194

errors, measurement 148–9

ethical issues

abortion 81–3

drugs 26, 34–7

ethics committees 38–9

genetic diseases 75, 80–6

genetic engineering 98–9

'natural' versus 'good' 99

principles 35–37

utilitarianism 36

ethics 35–7

Eudoxus, astronomer 194–5

eugenics 83

evidence, importance of 12, 165

evolution 101–14

artificial selection 106

evidence 103–4

fossils 103, 109

genetic basis 111–12

mechanism 101, 102, 105–8

mutations 101, 112

natural selection 101, 102, 105–8, 112–14

reactions to theory of 110–11

and religion 110–11

timescale 108–10

exercise, and heart disease 60

experimentation, importance of 200

explanations, scientific 1, 12

extra low frequency (ELF) radiation 186, 187–9

false negatives/positives, in testing 80

falsification 25

fats, heart disease 57–9

feedback, global warming 162

fermentation 6, 31

field 203

fish, genetically engineered 97–8

Fleming, Alexander 30–1

flu 21–2, 41–2

food chain, and radioactivity 133

food production, global warming impact 164

fossil fuels 116, 120, 121–2, 132–3

fossils 101, 103, 109

four stroke cycle 144

frameworks, scientific 12, 25

fuels

burning 144

efficiency 115, 118–19, 130–2

energy and 115–16

environmental issues 115–16, 121–8, 132–41, 144–54

fossil 116, 120, 121–2, 132–3, 144

and global environment 155–66

inequality 115

pollution 125–6, 144–5

sources 116, 120–2

using 115–28

future

energy use 126–8, 136–41

genetic engineering 95–8

Galapagos Islands, evolution and 103–4

galaxies 213, 214–16

Galileo, astronomer 200–2

GDP see Gross Domestic Product

generalisations, importance of 12

genes 51, 52

evolution 111–12

genetic diseases 75–6, 77

genetic engineering 88–9

model of inheritance 76

selfish 113

genetic basis for evolution 111–12

genetic counselling 82

genetic diseases 75–86

antenatal testing 78–82
    ethical issues 75, 80–6
genetic engineering 87–100
    agriculture 87, 89–95
    described 89
    environmental issues 92–5, 97–8
    enzymes 89
    ethical issues 98–9
    fish 97–8
    future 95–8
    genetic diseases 83
    hazards 97–8
    industrial aspects 87, 89–93
    medical research 96
    principles 88–9
    public opinion 99–100
    regulation 100
    safety 93–5
    transplants 96, 97
    vaccines 95
genetic modification (GM) see genetic
    engineering
genetic screening 84–6
geocentric models, astronomy 195, 197,
    201
geometric increase concept, and evolution
    105
germ theory of disease 1 12, 19
germs
    and diseases 1–12, 19
    identifying 7–11
giraffes, evolution and 107
global environment, fuels and 155–66
global trends, energy 120
global warming 127, 155, 158–66
GM (genetic modification) see genetic
    engineering
gravitation, universal 203–5, 219
gray (Gy), unit of radiation dose 168, 170
greenhouse effect 127, 155, 158–66
Gross Domestic Product (GDP), energy
    and 119, 128
Gy see gray

haemophilia 76, 83
half-life, radioactivity 173, 175
health
    defined 64
    diet and 14, 57–9
    global warming impact 164
    health screening 85–6

perceptions 63–4, 65
    pollution 149–50
    problems 63–4
    see also risks, health
heart disease 53–61, 71
heliocentric model, astronomy 196–8, 201, 205
herbal remedies 67–8
herbicides 92–3
Herschel, William 203–4, 213
Herzsprung-Russell diagram 211–2
HIV and AIDS 40–1
holistic medicine 64–74
homeopathy 67–8
Hubble, Edwin 214–15, 216–18
Huntington's disease 77–8

ideas, scientific see theories
imagination, importance of 1, 13, 192, 208, 214
immune system 1, 14–15, 16, 40
    allergies 67
immunisation 16–21
incidence, of disease 182
inductive reasoning 25
Industrial Revolution, air pollution 152–3
infant mortality 63
infection, resistance to 14
influenza 21–2, 41–2
inheritance, gene model 76
insulin, genetically engineered 89
interdependence, of species 94–5, 97–8
ionising radiation 167–71, 179
    cancer and 170–2, 181 5
    see also radiation
irradiation vs contamination 172–3
island universes 213, 214–15
isotopes 169

Jenner, Edward 16–17
joule (J), energy measurement unit
    117–18, 120

Kant, Immanuel, philosopher 213
Kepler, Johannes, astronomer 199–200
kilowatt-hour (kWh), electricity
    measurement unit 130, 137
knowledge, reliability of 1, 24–5
Koch, Robert 6–8, 18, 28
Kuhn, Thomas, historian 25
kWh see kilowatt-hour

Lamarck, Jean Baptiste de 106–7

Leavitt, Henrietta Swan 211–12, 213

legislation, air quality 151–4

leukemia 183–5, 189

Leverrier, Urbain, astronomer 204–5

life expectancy 63

light year, astronomical
  measurement unit 209, 210

lipoproteins, heart disease 58

low-level ionizing radiation 181–3

lung diseases 49–53, 176, 182–3

Magellanic Clouds 211–13

magic bullet concept, chemotherapy 28–9,
  31

magnetic fields, exposure to 588–9

maize
  genetically modified 91–3

Malthus, Thomas 105

Marshall, Barry 10–11

mass media 35, 46–7, 87, 100, 127, 179–81, 190

measurements, accuracy of scientific 148–9,
  160, 190

mechanism, importance in scientific explanations 62, 66,
  202–3

media reports, health risks 173, 180–1

medicine, alternatives in 62–74

medicines 26–42
  and drugs 27

Mendel, Gregor 111–12

miasma theory of disease 4–5

microbes, discovery 6–8

microscopes, diseases and 7–8

microwaves
  cosmic background radiation 208, 218–19
  mobile phones 185–7

MMR vaccine, autism 23–4

mobile phones 179, 185–7

molecular structure, drugs 32–3

multifactorial risks 60–1, 84

mutations
  cancer 51–3, 170, 182
  evolution 101, 112
  viral 22, 41–2

natural selection, evolution 101, 102,
  105–8, 112–14

'natural' versus 'good', ethical issues 99

nebulae 212–15

Neptune, discovery of 203–5

neutron stars 219

Newton, Isaac 202–3, 210

NFFO see Non-Fossil Fuel Obligation

nitrogen oxides 144–5, 149

Non-Fossil Fuel Obligation (NFFO) 137–8

nuclear fusion, in stars 219

nuclear power 116, 129–30, 131, 132–6, 167

nuclear waste 133–5

nuclides 169

open-mindedness, importance of 165–6

Organisation for Economic Co-operation and
  Development (OECD) 121

Origin of Species 107, 109, 110

ozone 145, 146, 149

paradigms, scientific 25

parallax 199
  stellar 197, 199, 209–11

particulates, pollution 143–50

Pasteur, Louis 6–8, 18–20

pasteurisation 15–16

patterns
  in data 12

penicillin 30–1

Penzias, Arno 218–9

periods, stellar 212

pharmaceutical industry 27–8, 72

phenylketonuria (PKU) 85

photography, astronomy and 210

photovoltaic cells 116, 130, 139–40

pigeons, and evolution 106

pigs, for human transplants 96–9

PKU see phenylketonuria

placebos 37–8

planets see solar system

plants, drugs source 27

PM_{10}s, pollution 147, 148, 149, 150

politics
  diseases 13, 14, 41
  electricity supplies 129, 132–42
  pollution 151–4

pollution
  air 125–6, 143–54
  politics 151–4

Popper, Karl, philosopher 25

power lines, electricity 187–9

predictions, importance of 12, 25
  astronomy 208–9, 216, 218–19
  solar system 192, 204–7

prontosil (drug) 29

prospective studies, epidemiology 56
proving a negative 7, 48
Ptolemy, astronomer 195–6
puerperal fever 2–4
pump, Broad Street 5–6

quality of life concept 125–6

rabies 20–1
radiation
    dose 168, 170, 173, 175–7
    effective dose 170–2
    from radioactive sources 171
    health risks 132–5, 167–91
    sources 173–6
    types 158–9, 167, 170–1
radioactivity 167–78
    movement of radioactive material 173–6
    radioactive waste 133–6
radon 173–6, 181–3
recessive genes 77
redshifts, astronomy 215–17
regulation, of science and technology 34, 37–8, 97,
    100, 151–4
Relativity, General Theory of 208, 216,
    218, 219
reliable knowledge, science as 1, 24–5
religion
    and astronomy 197–8, 200–2
    and evolution 110–11
renewable energy 116, 129, 136–41
reserves, energy 121–2
reservoirs, carbon 156–7
resistance
    drugs 26, 32, 40–2, 112
    infection 14
    to theories 3–4, 11, 13, 19, 196–7, 200–5, 217
retrospective studies, epidemiology 50
risks, health 43–61
    estimating 44–5
    expressing 45
    factors 46–9, 60–1
    interpreting 43, 48
    mobile phones 179, 185–7
    multifactorial risks 60–1, 84
    pollutants 149–50
    radiation 167–91
    reacting to 45–6
    smoking 49–53
salmon, genetically engineered 97–8

sampling, scientific 148–9
science
    challenging 62–74, 165–6
    commonsense and 192
    communication 190–1
    control concept 37–8, 47
    errors, measurement 148–9
    experimentation, importance of 200
    frameworks 12, 25
    open-mindedness 165–6
    process of 12, 25
    religion and 110–11, 197–8, 200–2
    sampling issues 148–9
    successful 24–5
    technology and 201
scientific community, role of 1, 25
screening, genetic 84–6
sea level changes, global warming impact 155,
    163
Seascale leukaemia cluster 183–5
selfish genes 113
Sellafield 133–5, 174, 183–5
Semmelweis, Ignaz 2–4
Shapley, Harlow 212–14
side-effects, drugs 26, 40, 65
sievert (Sv), unit of radiation equivalent dose
    170–2, 173, 176–7
singularity, cosmological 218
smallpox 16–17
smogs, photochemical 146, 153
smoking, health risks 49–53
Snow, John 4–6
solar cells 116, 130, 139–40
solar system 193
    understanding the 192–207
species
    interdependence of 97–8
    evolution of 105
spectroscopy, astronomy and 210–12
standard of living concept 115, 125–6
stars
    Cepheid variables 211–16
    classifying 211–12
    distance to 209–12
    life history 219
    parallax 197, 199, 209–11
    variable 211–16
'steady state' cosmological theory 217
stellar parallax 197, 199, 209–11
streptomycin 39–40

stress, heart disease and 59–60, 71

sulfur dioxide 145, 149

sulphonamide drugs 28–9

Sun

    composition 210, 219

    see also solar system

Sv *see* sievert

Tay-Sachs disease 85

TB *see* tuberculosis

technology, and science 7, 201

telescopes 201, 203–5, 210–11, 218

theories

    forming 1, 7, 12, 25

    resistance to 3–4, 11, 13, 19, 196–7, 200–5, 217

thrombosis 55

tortoises, evolution and 103–4

total primary energy supply (TPES) 120–1, 127–8

TPES *see* total primary energy supply

transplants, genetic engineering and 96, 97

transport

    energy for 124–5

    opinions 126

tuberculosis (TB) 7, 15–16, 21, 40–1

    new drugs 39–41

ulcers, causes of 10–11

uncertainty, in science 148–9, 190, 212–5

universe

    expanding 215–20

    structure 212–15

    understanding the 208–20

Uranus, discovery of 203

utilitarianism 36

vaccination 13–14, 16–21

    decisions 23–4

    genetically engineered vaccines 95

    policy 23–4

    and viruses 13

variable stars 211–16

viruses 8–10

    anti-viral drugs 41–2

    and bacteria 8–10

    leukemia 183–5, 189

    mutations 22, 41–2

    PERVS 97

    vaccination 13–14, 16–21

VOCs *see* volatile organic compounds

volatile organic compounds (VOCs),
    pollution 145, 146

voyage of the *Beagle* 102–4

Voyager interplanetary missions 205–7

Wallace, Alfred 107–8

water supply, diseases and 5–6

waves, renewable energy source 138–9

weather patterns, global warming impact 162–4

weed control 92–3

white blood cells 14–15, 28

WHO (World Health Organisation) 17, 20, 40–1, 145

wind, renewable energy source 138

Wilson, Robert 218–9

xenotransplantation 96, 97